T0203021

SpringerBriefs in Molecular Science

More information about this series at http://www.springer.com/series/8898

Jun-ichi Yoshida

Basics of Flow
Microreactor Synthesis

 Springer

Jun-ichi Yoshida
Kyoto University
Kyoto
Japan

ISSN 2191-5407   ISSN 2191-5415 (electronic)
SpringerBriefs in Molecular Science
ISBN 978-4-431-55512-4   ISBN 978-4-431-55513-1 (eBook)
DOI 10.1007/978-4-431-55513-1

Library of Congress Control Number: 2015936380

Springer Tokyo Heidelberg New York Dordrecht London

Printed on acid-free paper

Springer Japan KK is part of Springer Science+Business Media (www.springer.com)

Preface

This published work is intended to serve as a sourcebook on flow microreactor synthesis for students and researchers who are not familiar with this field. This book is aimed at developing a deeper understanding of chemical reactions by taking advantage of the characteristic features of flow microreactors. This book is not a compilation of all fields of flow microreactor synthesis, but it focuses on high-resolution reaction time control and fast micromixing. I do hope that the topics discussed in this book will be sufficient to interest a wide range of readers in this fascinating field of chemistry and to encourage them to try chemical synthesis using flow microreactors. On each topic, I usually focus on the works that have been done in my research group as examples simply because I know them more in detail than other works reported in the literature. I hope that these examples and the accompanying discussions will serve as a guide to how to use a flow microreactor for chemical synthesis. I thank the members of my research group, including Dr. Aiichiro Nagaki, who have performed extensive work to develop the chemistry described in this book. I also appreciate the financial support of a Grant-in-Aid for Scientific Research and NEDO projects to enable our research in this field.

This book is based on the lectures that I gave at the Micro Chemical Production Study Consortium at Kyoto University for several years. This book was first published in Japanese as the basics part of a book entitled *Furoh-, Maikuro-Gousei* (Flow Microreactor Synthesis), by Kagaku Dojin, Kyoto, which I edited in 2014. I thank Kagaku Dojin for allowing me to publish the English version. I also thank Ms. Yukiko Nakayama of U-English for translation to make a draft of the English version and Ms. Yoko Uekawa for checking the manuscript, although I am fully responsible for the final version. I added Chap. 8, and also I made some rearrangements and modifications of the contents in several chapters. Therefore, this book is not a simple English translation, but a revised edition of the Japanese version. For the benefit of the readers who are willing to study wider aspects of flow microreactor synthesis, a list of books and review articles is provided as an appendix.

February 2015   Jun-ichi Yoshida

Contents

Chapter 1
Departure from Flask Chemistry

Abstract Reactors used for chemical synthesis are categorized into batch reactors and flow reactors. In a batch reactor, the concentration of chemical species including starting materials and products changes as time goes. In a continuous flow reactor at steady state, however, the concentration of chemical species remains constant as time goes, but it differs at a different spatial location in the reactor. The reaction time for the batch reactor equates with the spatial location in the flow reactor. The time for which the reaction solution resides in the flow reactor is called the residence time. The average residence time (mean residence time) is determined by the cross-sectional area and the length of the reactor and the flow rate. Appropriate setting of the residence time is required for controlling the reaction in a flow reactor.

1.1 Introduction

Researchers at universities and industries in chemistry and related fields have experienced various chemical reactions from their school days up to the present. Most of such chemical reactions were performed in batch type reactors of centimeter size, such as flasks and beakers. Such batch macroreactors, however, are rather special reactors (special reaction environments) for chemical reactions in nature. Flasks and beakers, which are designed to fit for human hands, were developed for easy handling by humans. Reactors in living organisms, such as organelles, cells, and tissues are much smaller than flasks. In addition, substances in nature react while moving or flowing, rather than residing at a single location. We should keep in mind that there are many types of reactors which are different from flasks and beakers. The chemistry of flow microreactor synthesis requires such a broader perspective of chemical reactions and reactors, departing from the flask chemistry that researchers have long been accustomed to.

© The Author(s) 2015   1
J. Yoshida, *Basics of Flow Microreactor Synthesis*,
SpringerBriefs in Molecular Science, DOI 10.1007/978-4-431-55513-1_1

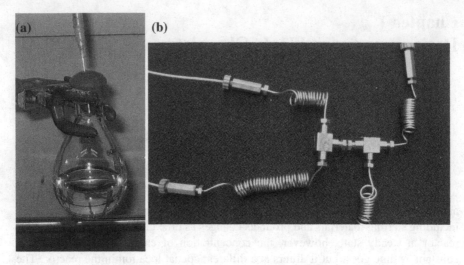

Fig. 1.1 Typical reactors for laboratory organic synthesis: **a** batch reactor, **b** flow reactor

1.2 Flow Reactors and Batch Reactors

Reactors for chemical reactions can be roughly categorized into two types: flow reactors and batch reactors (Fig. 1.1), although a variety of other reactors including semi-batch reactors are available. Organelles, cells, and tissues in living organisms are reactors with complicated structures and cannot be regarded as simple batch or flow rectors. A typical example of the flow reactors can be a chemical plant, where chemical products are mass-manufactured. In fact, most of the bulk chemicals such as methanol and acetic acid are manufactured using flow reactors in chemical plants. In contrast, many fine chemicals and drugs are manufactured using batch reactors.

The main theme of this book is flow microreactor synthesis, i.e., a chemical synthesis using a flow reactor having a microinternal space. Although researchers who have been accustomed to flasks and beakers can understand microreactors or smaller reactors relatively easily, they may find it difficult to understand flow reactors. Therefore, this chapter describes how reactions proceed in flow reactors.

1.3 Fluid Dynamics

A flow reaction proceeds in "motion" or in a "flow," whether it can be a flow of gas, a flow of liquid, or a flow of supercritical fluid. All such flows can be controlled by fluid dynamics. However, in this chapter we try to explain and understand the characteristics of flow reactions on the basis of reaction kinetics and synthetic organic chemistry, instead of explaining the phenomena on the basis of fluid dynamics.

1.4 Residence Time

The concept of residence time can be a major factor that hinders the understanding of reactions in flow reactors. Many flow reactors are tubular reactors, although there are many channel reactors as well. Let us consider how solution-phase reactions proceed in a tubular reactor, because a similar discussion can be made for channel reactors. A solution containing a starting material or materials is introduced at one end of the tube and undergoes reaction inside the tube. Then, the solution containing the product is discharged from the other end of the tube. The time for which the solution resides inside the tube is called the residence time, which equates with the reaction time, if a reaction occurs only in the tube. Although the average residence time or the mean residence time can be calculated easily by using the cross-sectional area and the length of the tube and the flow rate (Fig. 1.2), the calculation is not simple when the volume of the solution changes during the reaction. For example, when a gas evolves during the reaction, the calculation or determination of the residence time is difficult. Even if the reaction does not suffer from such a problem, there is another problem, i.e, the problem of residence time distribution. At a molecular level, some molecules move fast in the tube and other molecules move slowly. This causes the problem of residence time distribution. Also, on the macroscale, the linear flow velocity is lower near the tube wall and is higher around the center of the tube, and such a distribution of flow velocity also causes the problem of residence time distribution. Despite such a problem, the residence time is the key concept in flow chemistry and thus needs to be understood well. In this book, the discussions will be made based on the mean residence time unless otherwise noted, because there seems to be practically little effect of residence time distribution in most cases.

Fig. 1.2 The mean residence time

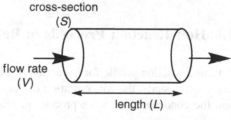

cross-section (S)

flow rate (V)

length (L)

mean residence time = (S x L) / V

1.5 Relationship Between Residence Time and Reaction Time

Setting the residence time is important for reactions using flow reactors. When the residence time is set too short, the solution can be discharged out of the reactor before the reaction is complete, leaving the starting materials unchanged. When the residence time is set too long, the solution meaninglessly continues to flow inside the reactor after the reaction is complete. The products can decompose during such excess time, if they are unstable under the reaction conditions. Therefore, the residence time must be set exactly the same as the time required by the reaction to complete.

The time required by the reaction to complete (the reaction time) can be regarded as the time required by most materials to react and convert into the products. The reaction time here intends to mean the time required by a group of molecules to react, and it is not the reaction time at a molecular level. The time required by a single molecule to react and convert into another molecule, or specifically the time required by a molecule of the starting material undergoes a reaction through a transition state to give a product molecule, can be in the order of several hundreds femtoseconds to picoseconds. This is substantially equal to the time for which a molecule oscillates. Even light can travel only 0.3 μm in a single femtosecond. Such times are much shorter than the reaction times that we experience. A reaction that requires an hour to take place would not mean that every molecule starts reacting coherently and requires an hour to react slowly into a product module. Each single molecule reacts in an extremely short time. When viewed as a group of molecules as a whole, the molecules do not react coherently but they react individually. Therefore, the reaction rate is determined by the number of molecules that react per unit time.

There is ambiguity in the word of reaction time, because it typically requires an enormous amount of time before all the starting material molecules are completely converted to the product molecules. It is difficult to define the reaction time on the macroscale precisely. What proportion of the starting material is to react to complete the reaction, 99 or 99.9 %? However, we will use the commonly accepted reaction time in which *most* of the starting material is consumed.[1]

1.6 How Reaction Proceeds in Batch Reactor

A typical reaction profile for a reaction in a batch reactor is shown in Fig. 1.3. Once the reaction starts, the concentration of the starting material in the reactor decreases and the concentration of the product increases with time. The reaction kinetics is

---

[1]Ahmed Zewail determined the reaction time at the molecular level by using femtosecond lasers and won the 1999 Nobel Prize in Chemistry for the work [1].

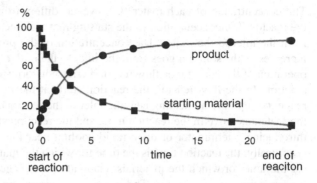

Fig. 1.3 Progress of a reaction in a batch reactor

discussed on the basis of such changes with time or temporal changes in the concentrations of the materials. In the first-order reaction, for example, the rate of the reaction, the rate of decrease in the concentration of the starting material is proportional to the concentration of the remaining starting material.

1.7 How Reaction Proceeds in Flow Reactor

A reaction in a flow reactor occurs basically in the same manner, but it needs a slightly different perspective of time. In a flow reactor, the concentration at each location in the reactor does not change with time but usually remains constant at steady state under continuous flow conditions (Fig. 1.4). In other words, the starting materials to react are fed into the inlet usually at a constant concentration and at a constant rate, and such a continuous flow condition makes the concentrations of all the materials involving the reaction constant at each location in the reactor.

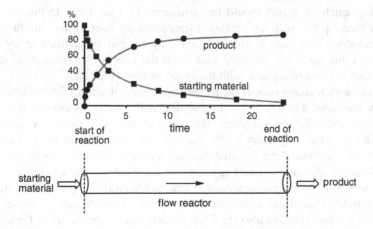

Fig. 1.4 Progress of a reaction in a flow reactor

The concentration of each material, however, differs from at different locations in the reactor. The concentration of the starting material decreases as the solution goes from the inlet to the outlet. The concentration of the product, in contrast, usually increases as the solution goes from the inlet to the outlet. Therefore, the start of the operation of the continuous flow reaction does not correspond to the start of a batch reaction. In the flow reactor, the reaction starts at the inlet (or upon introducing a reagent or upon rapid heating near the inlet of the reactor). The reaction proceeds as the materials move in the reaction tube, and the reaction stops at the outlet (or upon introducing a terminator or upon rapid cooling near the end of the reactor). More specifically, the reaction time is the time for which the materials reside in the reactor (or the time for which the materials reside after the reagent is introduced and until the quencher is introduced). The materials entering the inlet of the flow reactor move in the reactor at a certain rate and reach a certain point in a predetermined period of time. When the flow rate is constant, the time is proportional to the distance from the inlet. In other words, the flow reactor replaces the time with the space. The reaction time is zero at the inlet, and the reaction time increases toward to the outlet.

1.8 How to Set Residence Time

Setting the residence time for flow reactions is not very easy. The reaction needs to be started in an appropriate manner around the inlet, and the reaction needs to be terminated when the concentration of the starting material reaches almost zero and the concentration of the product becomes sufficiently high. You may assume that the reaction is terminated when the solution is discharged from the reactor. However, we must notice that the reaction can actually still proceed after the discharge from the reactor. In some cases, the reaction proceeds in the collecting flask! To terminate the reaction in a reliable manner, a reagent to terminate the reaction (quenching agent) should be introduced into the tube. In this case, the reaction time equates with the residence time from the inlet to the introduction of the terminator (vide supra). The reaction may instead be terminated by rapid cooling. In this case, the residence time from the inlet to the point at which the temperature is lowered equates with the reaction time (Fig. 1.5).

The discussion above reveals the prime importance of setting the residence time for flow reactions. The residence time setting will now be discussed. If the rate constant of the reaction is known, the reaction time can be easily calculated depending on what proportion of the starting material we want to be reacted to complete the reaction. The calculated reaction time can then be set as the residence time. However, the rate constant is unknown for most reactions that are used in organic synthesis. In organic reactions, a minor structural change in the molecules of the starting material can often greatly change the rate constant of the reaction. Moreover, the rate constant also significantly depends on the nature of the solvent. Predicting the reaction rate is thus very difficult. For flask reactions, even if we do

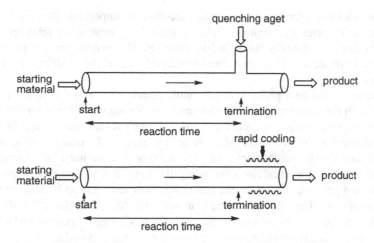

Fig. 1.5 Reaction time and methods for terminating a reaction

not know the reaction rate constant, we can monitor the reaction by determining the concentration of the starting material or the product during the course of the reaction, and the reaction can be terminated at optimum timing. However, flow reactions cannot be controlled in this way. For flow reactions, the reaction time can be determined only by carrying out reactions with various residence times and selecting the optimum residence time. The residence time can be changed simply by changing the flow rate, and the optimum residence time can thus be determined relatively easily by changing the flow rate (Fig. 1.6). However, we must notice that changing the flow rate can change the mixing rate and other conditions (see Chap. 3 for details). To change the residence time without changing the flow rate, reactor tubes with varying lengths should be prepared and the optimum residence time should be determined by changing the reactor (Fig. 1.6).

One common problem in shifting from batch reactors to flow reactors is the lengthy reaction time of conventional reactions that are often used in organic synthesis. Batch reactions can often require the reaction time of 1–24 h, but the reaction time or the residence time of 1–24 h is unrealistic for flow reactions.

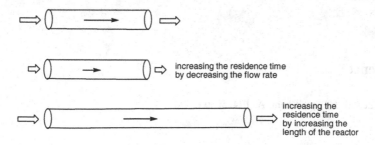

Fig. 1.6 How to change the residence time

For flow reactors, selecting relatively fast reactions is important. Reactions developed in laboratories of organic synthesis have been limited to relatively slow reactions that can be easily controlled in flasks. In other words, most conventional reactions have been discovered and developed under the restriction of flask environments.

However, it can be a problem if we simply compare the reaction time for batch reactors with the residence time for flow reactors. We see many cases reporting, for example, the reaction that requires an hour in a batch reactor can be completed in the residence time of a minute or so in a flow reactor. In reality, however, the reaction time for the batch reactor and for the flow reactor must be basically the same under the same conditions including the reagent concentration, equivalent ratio, and temperature. However, the reaction in flow reactors can be faster because flow reactors are often smaller in size than batch reactors. Therefore, we often call them flow microreactors. A smaller flow reactor has a larger specific surface area and thus can be heated rapidly so that the reaction would proceed at a rate intended at the temperature, whereas it takes time to increase the temperature of the entire container after the batch reactor starts being heated. Another cause for the fast reaction in flow reactors can be the rapid cooling. The rapid heat transfer by virtue of high surface-to-volume ratio enables the control of highly exothermic reactions, and the reaction can be performed at a natural rate. In contrast, in reactions in batch reactors the reagent needs to drop and mix slowly due to slow heat removal. Moreover, in two-phase reactions substances move at limited rates at interfaces in batch reactors, but the reactions in flow microreactors eliminate such limitations, because of a large interfacial area per unit volume. Other than the cases described above, the reaction rate is basically the same for flow reactors and batch reactors.

1.9 Looking Forward

Researchers will be freed from the limitations posed by the flask chemistry and will develop, with great anticipation, synthetic reactions that have been impossible with flasks, and will reveal the true values of flow reactions in the near future. Considerations needed in developing such flow reactions by taking advantage of characteristic features of flow microreactors will be discussed in the following chapters.

Reference

1. A.H. Zewail, J. Phys. Chem. A. **104**, 5660 (2000)

Chapter 2
Controlling Residence Time

Abstract For a batch reaction, the conditions within a reactor (the concentrations of materials and products) change with time. In contrast, for a flow reaction, the conditions within a reactor remain constant with time, but the conditions differ at different locations within the reactor. The reaction time for a flow reaction can be controlled by controlling the residence time in the flow reactor. Quench-flow method, in which a quenching reagent is added at the exit of the reactor to quantify the products with the residence time being varied, is effective for analyzing flow reactions. In a reaction involving an intermediate that easily decompose, a product can be obtained in a high yield only when a quencher is added in a time range during which a sufficient amount of the intermediate is produced and most of the intermediate is yet to decompose. Although a batch reactor can be used if the reaction is slow, a flow reactor is needed for a fast reaction (high-resolution reaction time control). Because the reaction rate depends on temperature, the temperature–residence time map, which can be obtained by the quench-flow method at different temperatures, is effective in optimizing the flow reaction conditions to obtain the product in a high yield.

2.1 Introduction

Chapter 1 describes the need to depart from flask reactions that researchers have long been accustomed to and move our focus on flow reactions to understand flow microreactor synthesis. This chapter describes one of the specific characteristics of flow reactions, i.e, residence time, and explores how such a characteristic can be used in controlling chemical synthesis.

A flow reaction is usually carried out at a steady state after some period of time passes from the initiation of the operation. At a specific location in the reactor, the concentrations of chemical species including starting materials, products, and reaction intermediates are temporally constant. However, the concentrations of the chemical species differ at different locations in the reactor. In contrast, the

J. Yoshida, *Basics of Flow Microreactor Synthesis*,
SpringerBriefs in Molecular Science, DOI 10.1007/978-4-431-55513-1_2

concentrations of chemical species in a batch reactor, such as a flask, are constant at any location in the reactor after the reaction is started, as long as the solution is stirred sufficiently and is kept under well-controlled temperatures. However, the concentrations of the chemical species in the batch reactor change with time. These characteristics of flow reactions and batch reactions are described in Chap. 1.

2.2 Monitoring the Progress of a Reaction in a Flow Reactor: Inline Analysis

A question now is how we can monitor the progress of a reaction occurring in a flow reactor. The reactor using a microchip may be entirely imaged to determine the concentrations of specific chemical species at various locations. However, imaging the microtube reactor is not so easy. Also, types of chemical species that can be imaged are rather limited, and the accuracy of the method depends on the nature of the chemical species that we want to quantify. The concentrations of chemical species at different locations in a flow reactor may also be measured by sensors or measuring devices installed at the different locations in the reactor (inline analysis). A simpler method is to measure the concentrations by inline analysis at the exit of the reactor.

An example of such inline analysis is shown in Fig. 2.1, in which a flow attenuated total reflection (ATR) infrared spectrometer is installed at the exit of the

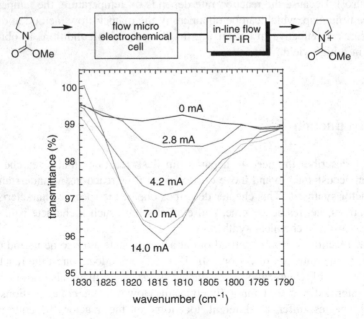

Fig. 2.1 An example of inline flow ATR analysis of an *N*-acyliminium ion generated by flow electrolysis [1]

reactor to analyze an N-acyliminium ion generated by anodic oxidation in the electrochemical flow reactor. The N-acyliminium ion shows a characteristic absorption at 1814 cm^{-1} due to carbonyl stretching vibration. The absorption intensity increases with the electric current, indicating that the concentration of the N-acyliminium ion increases with an increase in the current. Further increase in the current does not cause the increase in the concentration of the N-acyliminium ion. In this case, the concentration of the N-acyliminium ion can be determined by the inline flow ATR analysis.

2.3 Monitoring the Progress of a Reaction in a Flow Reactor: Quench-Flow Method [2, 3]

Inline analysis cannot be applied to quantitative analysis of all chemical species. Another method of monitoring the progress of a reaction is to add a quenching reagent (terminator) at the exit of the reactor and analyze the products with the residence time being varied by changing the length of the reactor or by changing the flow rate (quench-flow method). The products can be analyzed by conventional method for quantitative analysis such as gas chromatography or high-performance liquid chromatography, with which multiple chemical compounds can be analyzed at a time. Even in the case of a reactive species that is unstable and decomposes in a short time, a quenching reagent can instantaneously react with the species and convert it into a stable compound, which can then be analyzed by conventional methods to estimate the concentration of the active species.

2.4 Generation and Reactions of Oxiranyllithiums: An Example of Quench-Flow Method [4, 5]

As described above, reactions in flow reactors can be analyzed by the quench-flow method. In the example described below, you will see how effective the quench-flow method is for tracing the reaction and optimizing the reaction conditions.

Oxiranyllithium species are useful intermediates in organic synthesis and react with various electrophiles to give substituted epoxides [6–12]. The treatment of epoxides having an electron-withdrawing group such as an aryl group and a silyl group with s-BuLi leads to removal of a proton attached to the carbon bearing the group. This generates an oxiranyllithium species as shown in Fig. 2.2. The oxiranyllithium species is known to undergo decomposition accompanied by opening of the three-membered ring. Preventing such decomposition is crucial in the synthesis of substituted epoxides using oxiranyllithiums.

For example, deprotonation of styrene oxide 1 to generate oxiranyllithium 2 in the presence of TMEDA (tetramethylethylenediamine) followed by the reaction

Fig. 2.2 Generation and
reaction of oxiranyllthium

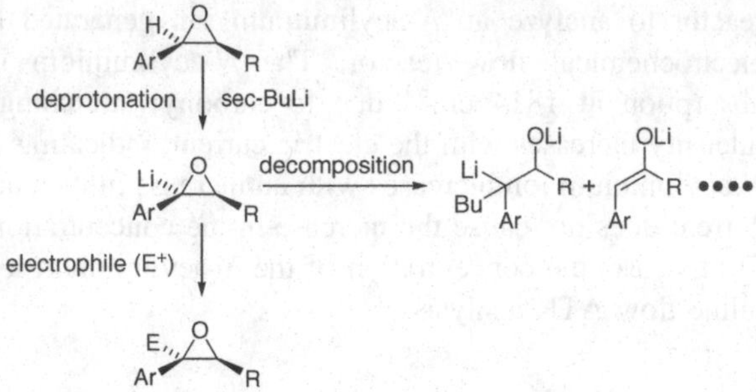

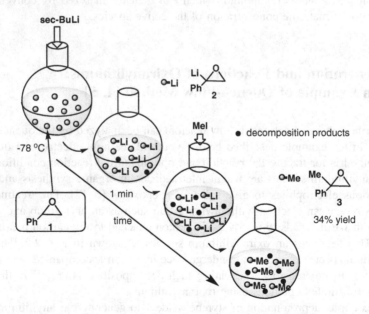

with methyl iodide at low temperatures such as −95 °C in a batch reactor gives the
desired compound **3** in a reasonable yield. However, the reaction at higher temperatures such as −78 °C in the absence of TMEDA gives rise to a significant
decrease in the yield (Fig. 2.3).

A typical procedure for this reaction in a batch reactor is as follows. A THF
solution of styrene oxide **1** was placed in a 20-mL round-bottom flask. While the
solution was being stirred with a magnetic stirrer in a dry ice cooling bath set at
−78 °C, a *s*-BuLi solution was added with stirring. After being kept at the same
temperature for a minute, a THF solution of methyl iodide was added. The desired
product **3** was obtained in only 34 % yield, suggesting that a large portion of the
oxiranyllithium species **2** decomposed during the time required by the dropping of

Fig. 2.3 Deprotonation of styrene oxide (**1**) to generate oxiranylltihium **2** followed by its reaction
with methyl iodide (MeI) in a flask at −78 °C

s-BuLi and the subsequent stirring to complete the conversion. To avoid such decomposition, the reaction should be carried out at lower temperatures in the presence of TMEDA.

The generation of oxiranyllithium **2** followed by its reaction with methyl iodide using a flow reactor can be carried out as follows (Fig. 2.4): A THF solution of styrene oxide **1** and a solution of s-BuLi are first mixed in a micromixer **M1**. Oxiranyllithium **2** is generated in microreactor **R1**. Subsequently, a THF solution of methyl iodine is added by using a micromixer **M2**. Oxiranyllithium **2** reacts with methyl iodide in a microreactor **R2** to produce the corresponding methylated product **3**, the concentration of which is determined by gas chromatography using an internal standard. The concentration of **2** is considered to be equal to that of **3** because we assume the reaction of **2** with methyl iodide is very fast. It is important to set an appropriate residence time in **R1**. The residence time should be sufficient for full conversion of **1** to **2**. However, the residence time should not be too long to allow decomposition of **2**.

The temperature is very important in controlling reactions because the rate of chemical reactions generally depends on the temperature. First, the reaction was conducted at −78 °C with the residence time being varied by changing the length of the reactor while keeping the flow rate constant. The amount of unchanged starting material **1** and that of product **3**, which were determined by gas chromatography, are plotted in Fig. 2.5 as a function of the residence time. Although these plots were obtained from separate experiments varying in the length (L) of the reactor R1, similar results must be obtained, in principle, for a single experiment using a long

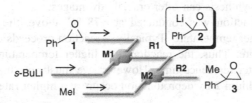

Fig. 2.4 Generation of oxiranyllthium **2** followed by its reaction with methyl iodide in a flow reactor

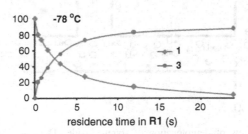

Fig. 2.5 Plots of recovery of **1** and yield of **3** against the residence time at −78 °C

flow reactor on which several analysis devices are attached at separate positions of the reactor. The recovery of **1** decreases and the yield of **3** increases with an increase in the residence time. In other words, the reaction proceeds with increasing residence time. Most of **1** is consumed in the residence time of about 25 s. This indicates that the reaction between styrene oxide (**1**) and *s*-BuLi is complete within the period of about 25 s.

The product **3** results from the reaction of oxiranyllithium **2** with methyl iodide, and thus, the content of the reactor is oxiranyllithium **2** before methyl iodide is added. More specifically, the molecules of the starting material **1** are deprotonated by *s*-BuLi while flowing in the reactor, and most of the starting material **1** is converted into oxiranyllithium **2** in 25 s. Figure 2.6 is a schematic view of this phenomenon. At −78°, almost no oxiranyllithium species **2** decomposed in the residence time of 25 s, which is said to be the time necessary for deprotonation. This phenomenon is explained by the results indicating that the amount of the starting material **1** decreases with increasing residence time and the amount of product **3** increases monotonously accordingly.

In this manner, a flow reaction can have a short residence time, allowing the oxiranyllithium species **2**, which is an unstable reactive species with a short lifetime, to react with methyl iodide before the species decomposes. In contrast, a batch reaction cannot be completed in a short reaction time due to the time for its conduction needed for the reaction. A batch reaction on an extremely small scale may be completed within a reaction time of 1 min or less, but a batch reaction on a large scale cannot be completed in a short time because the dropping of *s*-BuLi to a solution of **1** requires an additional time. Fast dropping in a batch reactor may cause rapid increase in the temperature which accelerates the decomposition of **2**. Thus, flow microreactor synthesis can offer crucial advantages.

Although the reaction was conducted at −78 °C above, the reaction is now conducted at a higher temperature. Typically, a reaction proceeds at a higher rate at a higher temperature. Thus, the reaction at a higher temperature is expected to complete in a shorter residence time. However, we must note that the reaction to produce oxiranyllithium **2** by deprotonation occurs at a higher rate as intended, but

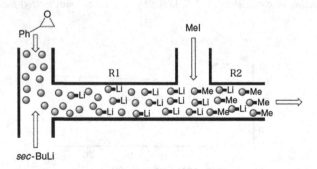

Fig. 2.6 Schematic view of deprotonation of styrene oxide (**1**) to generate oxiranyllthium **2** followed by its reaction with methyl iodide in a flow reactor

Fig. 2.7 Plots of recovery of **1** and yield of **3** against time at −48 °C

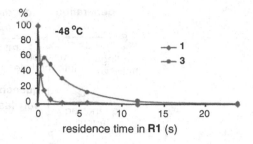

the oxiranyllithium also decomposes at a higher rate. Focusing on this, the results for a reaction at −48 °C (Fig. 2.7) will now be discussed.

At short residence times, the amount of the starting material or styrene oxide **1** decreases rapidly and the amount of the product **3** increases rapidly with the increasing residence time. After reaching its maximum yield of about 60 %, the amount of the product **3** decreases with the increasing residence time. The decreasing amount of the product **3** indicates that the intermediate oxiranyllithium **2** has decomposed in the reactor. The oxiranyllithium **2** starts to decompose before the starting material, styrene oxide **1**, is entirely deprotonated and converted into the oxiranyllithium **2**. In other words, the generation and the decomposition of oxiranyllithium **2** are not completely separable.

In contrast, at −78 °C, the generation and the decomposition of oxiranyllithium **2** are completely separable. The starting material **1** almost entirely converts into oxiranyllithium **2** before the oxiranyllithium **2** starts to decompose. Adding an electrophile thus allows the effective use of the oxiranyllithium **2** to obtain the desired product **3** in a high yield. However, this cannot occur at −48 °C, indicating that simply controlling the residence time would not necessarily control the reaction, because the reaction rate usually depends on temperature. It is thus important to control temperature as well as residence time to allow the reaction to proceed selectively to obtain the desired product in a high yield.

2.5 General Consideration on Controlling Reactions Involving Unstable Intermediates [13]

General consideration on controlling reactions will now be described. Let us focus on a reaction in which a starting material (**S**) undergoes reaction to generate an unstable intermediate (**I**) with time while **I** decomposes with time to give a by-product (**B**) (Fig. 2.8). When a sufficient amount of intermediate **I** accumulates, an appropriate quencher (**Q**) is to be added to obtain a desired product (**P**). We will now qualitatively explore the conditions under which a batch reaction can produce a desired product in a high yield, and the cases for which a flow reaction is needed by simple calculations based on the kinetics.

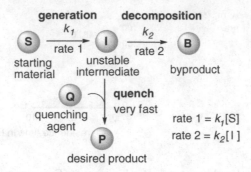

generation   decomposition

starting material
unstable intermediate
byproduct

quenching agent
quench very fast
desired product

rate 1 = k_1[S]
rate 2 = k_2[I]

Fig. 2.8 General scheme of the reaction involving an unstable intermediate (**I**) which is trapped by a quenching agent (**Q**) to give a desired product (**P**)

To simplify calculations, the generation and the decomposition of intermediate **I** are assumed to be first-order reactions, and the reaction of **I** with **Q** are assumed to occur at a sufficiently higher rate. Figure 2.9 shows temporal changes in the concentrations of starting material **S**, intermediate **I**, and by-product **B**, where the rate constant of generation reaction of **I** k_1 is 10 h^{-1}, which is 100 times the rate constant of decomposition of **I** k_2, and the initial concentration of **S** is 100.

The concentration of starting material **S** decreases with the increasing reaction time, and the concentration of intermediate **I** increases accordingly. In about 0.5 h, most of **S** is consumed and the concentration of **I** reaches its maximum. The concentration of **I** decreases with the further increase in the reaction time, because **I** decomposes into the by-product **B**. To obtain product **P** in a yield of 90 % or higher, quencher **Q** is to be added within the time range of about 0.25–0.6 h. This means that the reaction can be conducted in a flask without problems with its operations.

The situations change when the reaction is faster. A reaction at $k_1 = 10$ s^{-1} will now be discussed. The unit has been changed from hour to second, and thus, the reaction occurs at a rate that is 3600 times higher than the rate of the reaction described above. Although the concentration profiles shown in Fig. 2.10 are the

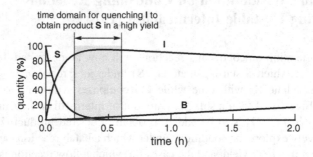

Fig. 2.9 Variation of the concentrations of starting material **S**, intermediate **I**, and by-product **B** with the reaction time for a slow reaction ($k_1 = 10$ h^{-1})

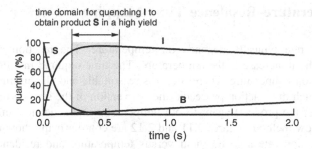

Fig. 2.10 Variation of the concentrations of starting material **S**, intermediate **I**, and by-product **B** with the reaction time for a fast reaction ($k_1 = 10 \text{ s}^{-1}$)

same, the reaction occurs in a different time unit, or specifically in seconds. In this case, product **P** can be obtained in a yield of 90 % or higher in the time range of about 0.25–0.6 s. This means that the reaction cannot be accomplished in a flask and that use of a flow microreactor is indispensable for such reactions. A flow reactor can set a reaction time precisely in an appropriate time range of the order of seconds or less and therefore allows product **P** to be obtained in a high yield. We call this high-resolution reaction time control. Notably, high-resolution reaction time control actually needs not only a flow reactor but also a microspace. We will discuss this in details in Chap. 3 and subsequent chapters.

However, the flow reactions also have limitations. If the generation and the decomposition of an intermediate occur at reaction rates substantially competing with each other, the maximum yield will be substantially low. This specifically occurs when k_1 and k_2 are substantially the same. In this case, controlling the residence time will not allow a product to be obtained in a high yield. One such example is the reaction of styrene oxide at −48 °C as described above. In this case, lowering the temperature may separate the generation and the decomposition of intermediate **I**, and product **P** can be obtained in a high yield. When decomposition of **I** is faster than the generation of **I**, the desired product **P** mostly cannot be obtained under any condition. In this way, the flow reactions are also limited by the reaction kinetics.

We should keep in mind that heat transfer is also a key to the success of controlling unstable short-lived intermediates. The oxiranyllithium species can be used efficiently in the flow reactor not only because the reaction time can be set short in the flow reactor but also because heat generated in the deprotonation can be removed quickly by taking advantage of fast heat transfer, which is unique to microreactors, rather than to flow reactors. An important factor is the small diameter of the reactor and the large surface area per unit volume. To allow smooth generation and reaction of the oxiranyllithium species, the reactor thus must be a flow microreactor, instead of simply being a flow reactor.

2.6 Temperature–Residence Time Mapping

The reaction rate typically depends on temperature. The reaction rate usually increases with an increase in the temperature. The time required in a flow reaction also depends on temperature. Therefore, it is reasonable and also useful to plot the degree by which the reaction proceeds (the conversion of the starting material and/ or the yield of the product) against both temperature and residence time based on the quench-flow method. Figures 2.11 and 2.12 are contour maps showing the plots of the conversion rate and the yield versus temperature and residence time for reactions in which oxiranyllithium is generated from styrene oxide and reacted with methyl iodide [4, 5]. The conversion of the starting material and the yield of the product are low at a lower temperature and a shorter residence time. The conversion and the yield increase with an increase in temperature and residence time. However, further increase in temperature and residence time causes a decrease in the yield because of decomposition of the oxyranyllithium intermediate. With this contour

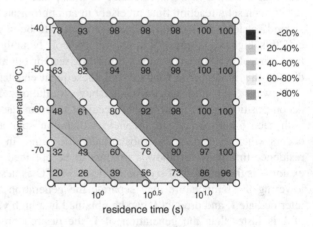

Fig. 2.11 Plots of the conversion of styrene oxide **1** against temperature and residence

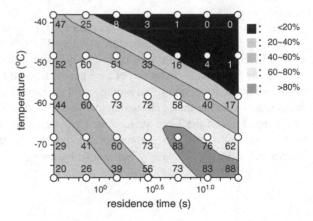

Fig. 2.12 Plots of the yield of methylated product **3** against temperature and residence

map, it is easy to determine the optimal temperature and residence time. Therefore, the temperature–residence time map serves as a powerful tool in analyzing flow reactions and optimizing the reaction conditions. Chapter 4 describes how to use the temperature–residence time map in detail.

2.7   Looking Forward

High-resolution reaction time control based on controlling the residence time discussed in this chapter is one of the most important features of flow microreactor synthesis. This feature enables chemical synthesis that cannot be done in batch. Some examples will be discussed in the following chapters.

References

1. S. Suga, M. Okajima, K. Fujiwara, J. Yoshida, J. Am. Chem. Soc. **123**, 7941 (2001)
2. T.E. Barman, S.R.W. Bellamy, H. Gutfreund, S.E. Halford, C. Lionne, Cell. Mol. Life Sci. **63**, 2571 (2006)
3. W.P. Bula, W. Verboom, D.N. Reinhoudt, J.G.E. Gardeniers, Lab. Chip. **7**, 1717 (2007)
4. A. Nagaki, E. Takizawa, J. Yoshida, J. Am. Chem. Soc. **131**, 1654, 3787 (2009)
5. A. Nagaki, E. Takizawa. J. Yoshida, Chem. Eur. J. **16**, 14149 (2010)
6. T. Satoh, Chem. Rev. **96**, 3303 (1996)
7. G. Boche, J.C.W. Lohrenz, Chem. Rev. **101**, 697 (2001)
8. S. Florio, Tetrahedron **59**, 9693 (2003)
9. V. Capriati, R. Favia, S. Florio, R. Luisi, ARKIVOC, xiv, 77 (2003)
10. D.M. Hodgson, C.D. Bray, P.G. Humphreys, Synlett **1**, 0001 (2006)
11. D.M. Hodgson, P.G. Humphreys, S.P. Hughes, Pure Appl. Chem. **79**, 269 (2007)
12. V. Capriati, S. Florio, R. Luisi, Chem. Rev. **108**, 1918 (2008)
13. J. Yoshida, Chem. Rec. **10**, 332 (2010)

Chapter 3
Fast Micromixing for High-Resolution Reaction Time Control

Abstract To allow residence time control in a flow reactor, the mixing needs to be complete in a time shorter than the residence time. Mixing is eventually caused by molecular diffusion. The time required by molecular diffusion most greatly affects the mixing time, and the time required by molecular diffusion increases in proportion to the square of the diffusion distance. To allow fast mixing, the diffusion distance needs to be set short, but shortening the diffusion distance by mixing caused by stirring has limitations. A more effective way to shorten the diffusion distance is to make small solution segments by using a microstructure. A system consisting of a micromixer and a flow microreactor is effective in setting a short residence time and thus is essential for conducting extremely fast reactions in a controlled way (high-resolution reaction time control).

3.1 Introduction

Chapter 2 focuses on the importance of what we call the residence time in flow reactions, and we have learned that controlling the residence time allows us to selectively obtain a desired product even if a product or an intermediate can decompose quickly under the reaction conditions. Concerning the validity of the residence time control, however, a reaction with a long reaction time does not need a flow reactor, but it can be carried out simply by using a conventional batch reactor. For a reaction that takes an hour or a day, the reactants are put in a batch reactor to start the reaction, after which there is a sufficient amount of time before the reaction is terminated by adding a quenching agent to the reactor.

A flow reactor is, however, inevitable for conducting a fast reaction that requires setting of a short residence time. In Chap. 2, we have learned that the residence time (reaction time) should be set in a narrow range to obtain the desired product in high selectivity (high-resolution reaction time control). In setting a short residence time for a flow reactor, we must note the following. Many reactions are started by mixing

© The Author(s) 2015

J. Yoshida, *Basics of Flow Microreactor Synthesis*,
SpringerBriefs in Molecular Science, DOI 10.1007/978-4-431-55513-1_3

two or more substances, such as a starting material and a reagent or a catalyst, although there are other ways to start reactions such as heating and irradiating light.

Let us consider a reaction started by mixing two substances in a flow reactor and terminated by introducing a quenching agent into the reactor after a residence time of one second. The reaction time is one second. To guarantee the reaction time of one second, the time taken by the two substances to mix into a homogeneous solution (mixing time) needs to be sufficiently shorter than one second. If the mixing is slow and takes one second or so, the reaction time is indefinite. In other words, the reaction kinetics cannot be applied within the residence time. If the solution is not homogeneous in the flow reactor, the reaction control based on the kinetics is completely meaningless. Thus, controlling the residence time in a flow reactor is valid only when the mixing time is sufficiently shorter than the residence time.

However, this is difficult to get fast mixing in a conventional batch reactor and also in a large flow reactor, raising the need for a flow microreactor. You may remember that the generation and the reaction of oxiranyllithium species described in Chap. 2 would fail if the residence time cannot be controlled in the order of seconds. This is why a flow microreactor equipped with a micromixer is needed for this reaction. This chapter focuses on extremely fast mixing in a micromixer, which is needed to set a short residence time.

3.2 What Is Mixing?

Mixing, or a process in which two solutions mix into a homogeneous solution, will now be discussed. Let us assume that two solutions (solution A and solution B) are placed separately on both sides in a partitioned container, and the partition is removed promptly without disturbing the solutions (Fig. 3.1). No stirring is performed. In this case, the two solutions mix only by molecular diffusion, and eventually a homogeneous solution forms. Molecular diffusion is a phenomenon in which molecules move randomly by thermal motion, causing molecules concentrated at one location to diffuse into a larger area with time.

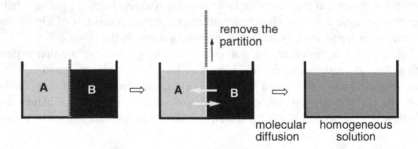

Fig. 3.1 Mixing of two solutions by molecular diffusion

Table 3.1 Rough correlation between the diffusion distance and the diffusion time for small molecules in water

Diffusion distance (µm)	Diffusion time (s)	Note
1	0.0005	The size of a bacterium
10	0.05	The size of an animal cell
100	5	
1000 (1 mm)	500 (8.3 min)	
10,000 (1 cm)	50,000 (14 h)	

Molecular diffusion can occur in a solid, a liquid, a gas, or a supercritical fluid. The theory of molecular diffusion is derived from the theory of Brownian motion developed by A. Einstein in 1905. The theory of Brownian motion is based on the following principle. The molecules of a liquid in thermal motion collide with particles (Brownian particles) suspended in the liquid, giving energy to the particles. Large particles will undergo collisions in all directions, and thus, the energy is balanced, whereas small particles will receive instantaneous energy that is not balanced, with the magnitude and direction of the energy varying randomly. As a result, the particles move randomly. The theory was later experimentally proved by J. B. Perrin.

The time required by molecular diffusion is proportional to the square of the distance of diffusion (diffusion distance or diffusion path length). Rough correlation between the diffusion distance and the diffusion time for small molecules in water is shown in Table 3.1. When the diffusion distance is 1 µm, which is the size of a bacterium, the diffusion time is 0.0005 s (0.5 ms). When the diffusion distance is 10 µm, which is the size of an animal cell, the diffusion time is 0.05 s. It is interesting to know that substances can diffuse in an extremely short time in an animal cell. For the diffusion distance of 100 µm, however, the diffusion time would be relatively long as perceived by humans. For the diffusion distance of 1 mm, the time required by the diffusion is in the order of minutes, and for the diffusion distance of 1 cm, the time required by the diffusion is in the order of hours. Therefore, the average diffusion time is larger, and thus, the mixing requires a longer time as the lateral size of the container is larger. A normal reactor, such as a flask, has a size larger than 1 cm and thus would require an extremely long time for mixing by diffusion alone. This raises the need for stirring.

3.3 Mixing by Stirring [1, 2]

Let us discuss how mixing is caused by stirring. Figure 3.2 shows a device commonly used for organic synthesis experiments in a laboratory. On this device, solution A is in a round-bottom flask, and solution B is added from the dropping funnel to the solution A, while the flask is being stirred using a magnetic stirrer. The reaction proceeds in the manner described below.

Fig. 3.2 A device consisting
of a flask, a dropping funnel,
and a magnetic stirrer

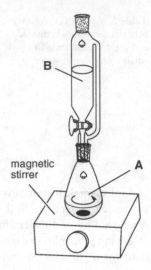

Figure 3.3 schematically shows typical mixing caused by stirring. First, stirring
generates eddies. The size of the eddy depends on the strength of the stirring. The
minimum size of the eddy is reported to be about 10–100 μm; no matter how
strongly the stirring is performed. Subsequently, the eddies diffuse. Then, molecular
diffusion takes place to eventually eliminate the eddies, resulting in a homogeneous

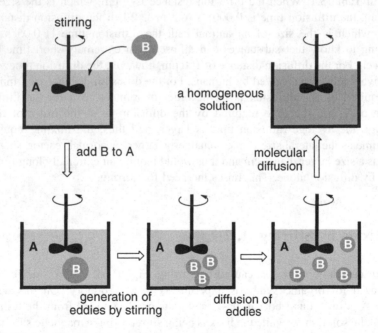

Fig. 3.3 Mixing caused by stirring in a batch reactor

solution. In the series of processes for mixing, the molecular diffusion requires the longest time.

As mentioned above, the time required by the molecular diffusion is proportional to the square of the diffusion distance. In this case, the diffusion distance corresponds to the size of the eddy. The strongest stirring would not reduce the size of the eddy to less than 10 μm. Relatively inefficient stirring, such as stirring using a magnetic stirrer, might generate eddies of about 100 μm. In this case, the time taken to achieve a homogeneous solution by molecular diffusion would be in the order of seconds.

On the basis of such a qualitative analysis or a thought experiment, a batch macroreactor commonly used for synthetic reactions in a laboratory, such as a flask, requires the mixing time in the order of seconds. In this case, controlling the reaction time in the order of seconds would be impossible in principle. The diffusion distance can also be a problem with a flow reactor when the reactor is large.

3.4 Mixing Using Micromixer [1, 2]

Let us consider mixing two solutions with a flow reactor. First, we assume one simple example in which two pipes are arranged to meet each other. In these pipes each having a large diameter, the flow is turbulent. However, the solution segments may not be very small, as in the mixing caused by stirring in a batch reactor. Thus, the diffusion distance is large, and the mixing is estimated to take the time in the order of seconds.

Now, we consider mixing using a smaller flow reactor. One typical example is a Y-shaped reactor shown in Fig. 3.4. In this reactor, laminar flow should be dominant. When the width of the flow passage is in the order of 100 μm, the time taken for diffusion would be in the order of seconds. In this case, fast mixing cannot be expected. When the width of the flow passage is in the order of 10 μm or less, the time taken for diffusion would be in the order of milliseconds. Such small tubes or channels, however, are not practically useful for synthesis because of only a small

Fig. 3.4 Mixing in a Y-shaped flow reactor

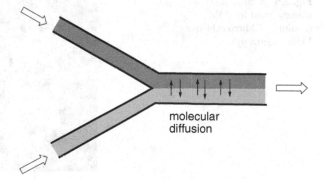

molecular diffusion

quantity of compounds can be produced. Also, such small tubes or channels would suffer from the problem of clogging.

More active and faster mixing can be achieved by using a micromixer. Whereas micromixers can be in various types with different principles, one of the most commonly used types has a microstructure to split a solution into small laminar flows, which are then recombined (multilamination-type micromixer). Figure 3.5 schematically shows this type of micromixer. This micromixer can reduce the size of one solution segment to the order of 10 μm or less, reducing the diffusion distance and enabling the mixing in a shorter time. The most known multilamination-type micromixer is a mixer fabricated at Institute of Microtechnique Mainz in Germany (Fig. 3.6) [3].

As shown in Fig. 3.7, another type of micromixers can shorten the diffusion distance further by contracting the flow in its width direction after the recombination (contracted flow) [4]. Further, split- and recombination-type micromixer split a liquid into multiple segments by using a microstructure and then recombine these segments, and then split the flow in a different direction and recombine the segments. Repeating such splitting and recombining eventually splits the liquid into

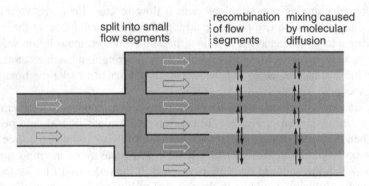

Fig. 3.5 Mixing in a multilamination-type micromixer

Fig. 3.6 A micromixer
manufactured by IMM
(Institute of Microtechnique
Mainz, Germany)

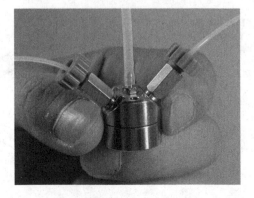

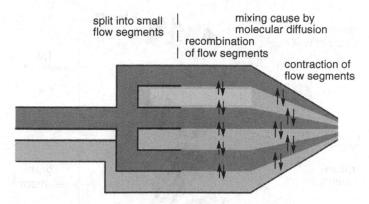

Fig. 3.7 Mixing based on multilamination and contraction of flow segments

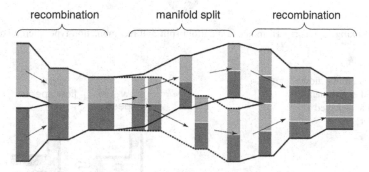

Fig. 3.8 Mixing based on manifold split and recombination of flow segments

small segments. This shortens the diffusion distance and thus enables faster mixing (Fig. 3.8).

Figure 3.9 shows the structure of a Toray Hi-mixer, [5] which is one example of such a split- and recombination-type micromixer. This mixer divides the solution into four at a time, and thus, the two layers introduced initially are split into eight layers at the first element and then into 32 layers at the second element, making the diffusion distance short very quickly.

Also, a simple T-shaped mixer can be effective for fast mixing when the flow speed is high. Microturbulence produced by engulfment flow [6] makes small segments of liquid and the diffusion distance becomes small, and allows faster mixing (Fig. 3.10). However, when the flow speed is low, laminar flow regime is predominant, and therefore, mixing should be slow. Figure 3.11 shows a transparent plastic model of a T-shaped micromixer.

As described above, the use of micromixers enables mixing in the order of milliseconds. A system consisting of a micromixer and a flow microreactor would enable a reaction with a residence time in the order of seconds or less. A macroflow

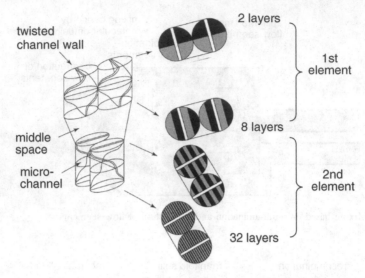

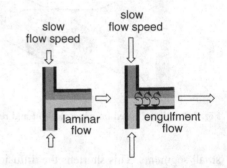

Fig. 3.9 Mixing in a Toray Hi-mixer (a manifold split and recombination-type micromixer)

Fig. 3.10 Laminar flow regime and engulfment flow regime in a T-shaped micromixer

Fig. 3.11 A transparent plastic model of a T-shaped micromixer manufactured by Sankoh-seiki

extremely fast reaction

$$H^+ + CH_3CO_2^- \longrightarrow CH_3CO_2H$$

fast reaction but slower than the above reaction

$$6H^+ + 5I^- + IO_3^- \longrightarrow 3I_2 + 3H_2O$$

Fig. 3.12 The Dushman reaction

system would not allow a reaction with such a residence time. Therefore, *micro* is essential for conducting extremely fast reactions that are complete in the order of seconds or less in a controlled way.

3.5 Measuring Mixing Efficiency or Speed

One typical method for measuring mixing efficiency uses the Dushman reaction [7], which involves competing two reactions, i.e., an extremely fast reaction and a reaction slightly slower than the fast reaction (Fig. 3.12). In detail, UV absorption of a solution produced by mixing a solution of hydrochloric acid and a solution of I^-, IO_3^{3-}, and CH_3CO_2Na is measured. When the mixing is fast and promptly generates a homogeneous solution, only the extremely fast reaction (protonation of acetate anion) proceeds selectively. When the mixing is slow, the concentrations of the reaction components are not uniform, causing the slower reaction to proceed as well, and generate iodine (I_2). The resultant solution shows absorption at 352 nm based on iodine. Thus, the mixing efficiency or speed can be estimated by measuring the UV absorption and determining the quantity of iodine.

3.6 Looking Forward

In this chapter, we have learned that the marked shortening of the diffusion path in a micromixer results in a mixing speed unobtainable by stirring in a macroreactor. The fast micromixing enables high-resolution reaction time control discussed in Chap. 2. Its applications to the use of short-lived reactive intermediates and related topics will be discussed in the following chapters. The fast micromixing also plays a crucial role in controlling extremely fast competitive consecutive reactions, which will be discussed in Chap. 7.

References

1. P. Rys, Acc. Chem. Res. **10**, 345 (1976)
2. P. Rys, Angew. Chem., Int. Ed. Engl. **12**, 807 (1977)
3. W. Ehrfeld, K. Golbig, V. Hessel, H. Löwe, T. Richter, Ind. Eng. Chem. Res. **38**, 1075 (1999)
4. P. Löb, K.S. Drese, V. Hessel, S. Hardt, C. Hofmann, H. Löwe, R. Schenk, F. Schönfeld, B. Werner, Chem. Eng. Technol. **27**, 340 (2004)
5. H. Wakami, J. Yoshida, Org. Process Res. Dev. **9**, 787 (2005)
6. A. Soleymani, H. Yousefi, I. Turunen, I. Chem, Eng. Sci. **63**, 5291 (2008)
7. S. Panić, S. Loebbecke, T. Tuercke, J. Antes, D. Bošković, Chem. Eng. J. **101**, 409 (2004)

Chapter 4
Use of Short-Lived Reactive Species Achieved by High-Resolution Reaction Time Control

Abstract Setting a short residence time in a precisely controlled manner using the flow microreactor system consisting of micromixers and a flow microreactor allows short-lived reactive species, which are considered so unstable that can be handled only at extremely low temperatures, to be usable before the species decompose. The temperature–residence time maps are highly useful in analyzing the reactions and optimizing the reaction conditions. Such approaches enable the generation of aryllithium species bearing ester carbonyl groups and allow them to react with subsequently added electrophiles.

4.1 Introduction

In Chap. 3, we learned that the flow microreactor system, consisting of micromixers and flow microreactors, enables reactions with residence times in the order of seconds or less. This chapter describes the principle of residence time control for reactions that involve short-lived intermediates. In particular, we focus on the reactions involving aryllithium species containing alkoxycarbonyl groups as intermediates.

4.2 Generation of Organometallic Species Containing Alkoxycarbonyl Groups

Halogen/lithium exchange, as well as hydrogen/lithium exchange, is widely used in organic synthesis to generate organolithium species. In particular, the halogen/lithium exchange reaction of aryl halides is often used to generate aryllithium species. Reagents commonly used for this reaction include butyllithiums such as n-BuLi, s-BuLi, and t-BuLi. Aryl halides commonly used for this reaction include aryl iodides and bromides. The resultant aryllithium species can be used to react

© The Author(s) 2015
J. Yoshida, *Basics of Flow Microreactor Synthesis*,
SpringerBriefs in Molecular Science, DOI 10.1007/978-4-431-55513-1_4

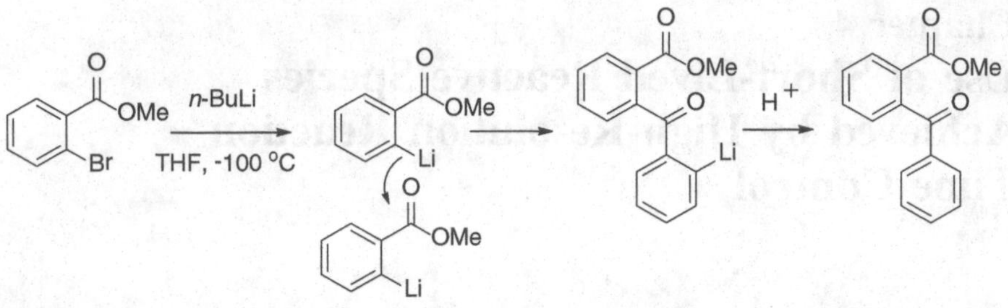

Fig. 4.1 Br–Li exchange reaction of methyl *o*-bromobenzoate with *n*-BuLi in batch

with various electrophiles including carbonyl compounds. It should be noted that reactions of aryllithium species with carbonyl compounds such as esters are usually very fast.

The Br–Li exchange reactions of alkyl *o*-bromobenzoates are known to form a dimer in THF (Fig. 4.1) [1, 2]. Presumably, the generated aryllithium species attacks the ester carbonyl group of another aryllithium species to generate an dimeric organolithium species, which then undergoes protonation. This makes it difficult to use the initially generated aryllithium species bearing the ester carbonyl group to react with subsequently added electrophiles at will.

In this way, the traditional chemistry assumes that aryllithium species and esters are not compatible with each other. The question now is whether this common knowledge in the current organic chemistry is really true. This question will be answered in the later part of this chapter.

4.3 Decreasing the Reactivity of Organometallic Reagents to Increase Compatibility

Organolithium compounds or Grignard reagents are highly nucleophilic among other organometallic reagents. Lowering the activity of an organometallic reagent will slow its reaction with an electrophile such as a carbonyl compound, thus allowing their compatibility. Typically, as the difference in electronegativity between metal and carbon is larger, the degree of polarization is larger, and accordingly, the reactivity of the organometallic reagent would be higher (Fig. 4.2). Therefore, organolithium species are generally the most reactive.

Extensive studies have recently been done to optimize the reactivity of organometallic compounds to make them compatible with functional groups such as carbonyl groups (Fig. 4.3) [3]. For example, an aromatic iodine compound containing an alkoxycarbonyl group undergoes an I–Zn exchange reaction in THF at −78 °C using $(CH_3)_3ZnLi$ to generate an aromatic zincate, which can then react

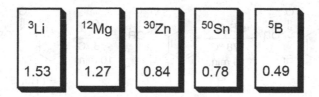

Fig. 4.2 The difference in electronegativity between metal and carbon

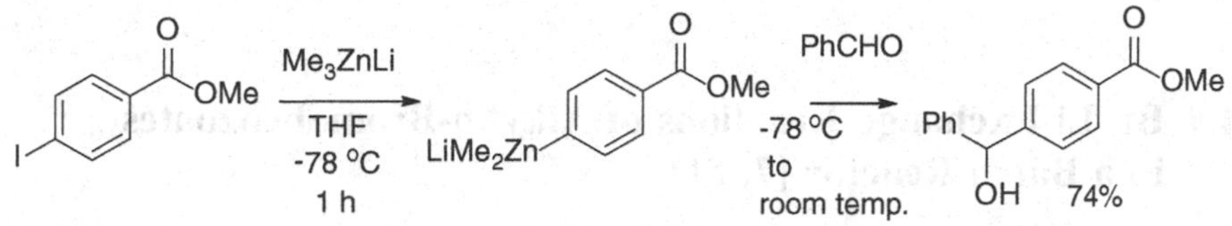

Fig. 4.3 Generation and reaction of an arylzincate bearing an alkoxycarbonyl group

with an electrophile [4, 5]. Arylzincates are, in general, much less reactive than aryllithium compounds allowing its coexistence with an ester carbonyl group. In another example, a diarylzinc species may be generated and undergo similar transformation. However, the halogen/zinc exchange needs a catalyst such as lithium acetylacetonate [Li (acac)] (Fig. 4.4) [6].

Although such less reactive organometallic species can be compatible with ester carbonyl groups and some other electrophilic functions groups, the reactions are slower than when organolithium compounds or Grignard reagents are used. A commonly used approach in the traditional chemistry has been to lower the reactivity of a reagent to decrease the reaction rate and to improve the selectivity as switching from organolithiums to organozincs. However, the flow microreactor synthesis can take a completely different approach. Instead of decreasing the reaction rate, this approach sets a short reaction time in a precisely controlled manner to maximize the selectivity of the reaction. This approach using the flow microreactor system will be described in detail in the following section.

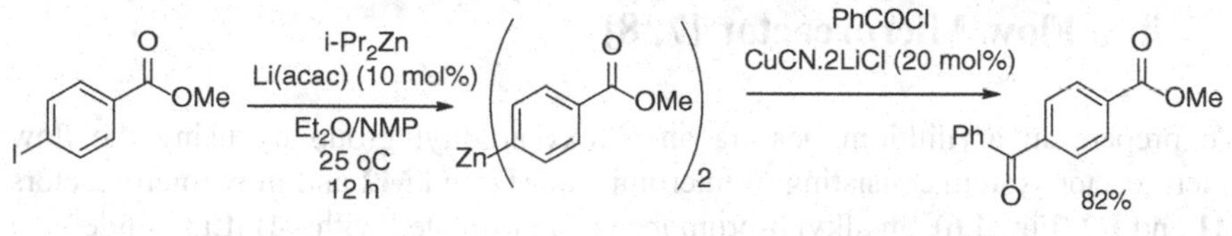

Fig. 4.4 Generation and reaction of a diarylzinc bearing an alkoxycarbonyl group

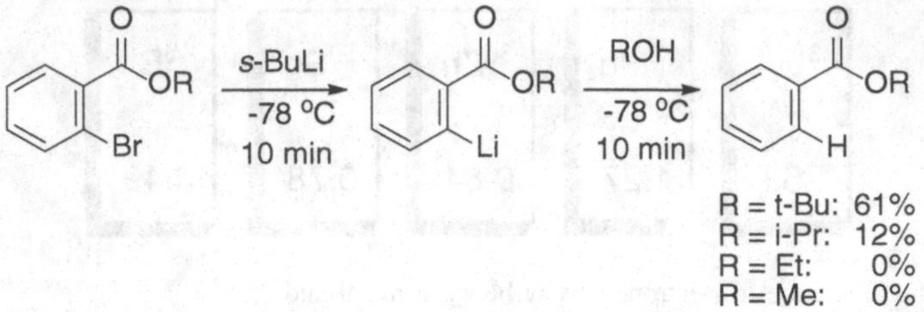

Fig. 4.5 Br–Li exchange reaction of alkyl *o*-bromobenzoates using *s*-BuLi followed by protonation in batch

4.4 Br–Li Exchange Reactions of Alkyl *o*-Bromobenzoates in a Batch Reactor [7, 8]

Generation and reaction of an aryllithium bearing an alkoxycarbonyl group will be discussed again. As mentioned earlier in this chapter, it is common knowledge that an ester and an organolithium reagent cannot be compatible with each other. In fact, aryllithium compounds bearing alkoxycarbonyl groups are difficult to prepare in a batch reactor from alkyl *o*-bromobenzoates using *s*-BuLi. In this case, *s*-BuLi is used, because it is electronically more reactive for Br–Li exchange reactions than *n*-BuLi, and is expected to be less reactive toward the carbonyl group because of the steric effects (Fig. 4.5).

For a methyl ester and an ethyl ester, for example, the treatment of the corresponding aryl bromide followed by quenching with an alcohol with an intention to trap the aryllithium intermediate would fail, thus producing no corresponding protonated product. This is presumably because the resultant aryllithium species reacts with the carbonyl group of another aryllithium species. For isopropyl ester, this process generates a small amount of protonated product. For *t*-butyl ester, the steric effects are so significant that the carbonyl carbon is less susceptible to nucleophilic attack, and thus, this process generates a protonated product, although the yield is about 60 %.

4.5 Br–Li Exchange Reactions of Alkyl *o*-Bromobenzoates in a Flow Microreactor [7, 8]

To prepare an aryllithium bearing an alkoxycarbonyl group by using the flow microreactor system consisting of micromixers **M1** and **M2** and flow microreactors **R1** and **R2** (Fig. 4.6), an alkyl *o*-bromobenzoate is treated with *s*-BuLi to undergo a Br–Li exchange reaction. Controlling the residence time for this reaction will allow generation of an aryllithium species as intended. The generated organolithium species then only needs to react with an electrophile before it decomposes.

Fig. 4.6 Br–Li exchange
reaction of alkyl o-
bromobenzoates using s-BuLi
followed by the reaction with
an electrophile using a flow
microreactor system

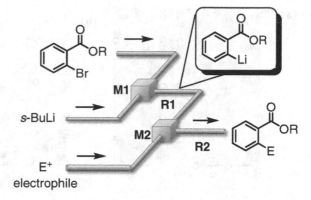

For example, the Br–Li exchange reaction of ethyl o-bromobenzoate was carried out using s-BuLi at −48 °C, and the resulting aryllithium species was then reacted with ethanol to give a protonated product. The results are shown in Fig. 4.7. The starting material, ethyl o-bromobenzoate, is consumed almost entirely within the residence time (t^{R1}) of 0.1 s, generating the protonated product in about 90 % yield. The yield of the product is lower at a longer residence time, presumably because the aryllithium intermediate should decompose. The optimal residence time at this temperature is ca. 0.1 s.

The rate of Br/Li exchange reaction, as well as that of the decomposition of the aryllithium, is dependent on temperature. The temperature–residence time map is thus useful in analyzing the reactions and optimizing the reaction conditions. In addition to the temperature–residence time map for an ethyl ester, those for other esters are also prepared (Fig. 4.8).

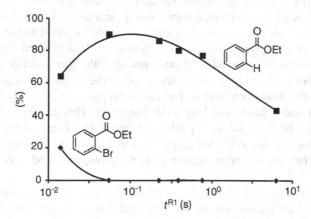

Fig. 4.7 Br–Li exchange of ethyl o-bromobenzoate followed by protonation with ethanol at −48 °C. Plots of the yield of the protonated product and that of the unchanged starting material against logarithm of the residence time

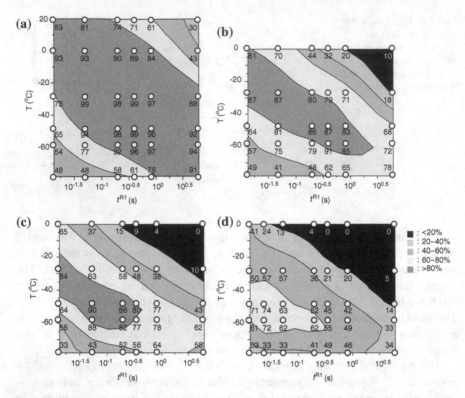

Fig. 4.8 Temperature–residence time map for Br–Li exchange of alkyl *o*-bromobenzoate followed by protonation, **a** t-butyl *o*-bromobenzoate, **b** isopropyl *o*-bromobenzoate, **c** ethyl *o*-bromobenzoate, and **d** methyl *o*-bromobenzoate

For the *t*-butyl ester, the protonated product is generated in high yields in a wide range of residence times and temperatures, except that the yield is low in a range of low temperatures and short residence times (at −60 °C or lower and at a residence time of 0.1 s or less), where the Br–Li exchange reaction proceeds insufficiently. Although about 60 % achieved by a batch reaction, the flow reaction allows substantially higher yields, almost quantitative yield of the product. This highlights the advantage of the flow microreactor. For the isopropyl ester, the yield is low in a range of high temperatures and long residence times. This can be explained by a dimmer formed by the resultant aryllithium reagent. However, the product is still generated in high yields in a wide range. This indicates the advantage of the flow microreactor, because the batch reaction gives the protonated product only in a very low yield (12 %).

For the ethyl ester, the product is generated in high yields in a narrower range, but the yield of the product can be 90 % or higher in one range. For methyl ester, the yield of 90 % or higher is not achievable in any range. However, it is amazing that the product is obtained in reasonable yields in a range of temperatures of −40 to −60 °C and residence times of 0.1 s or less, making clear contrast with a batch

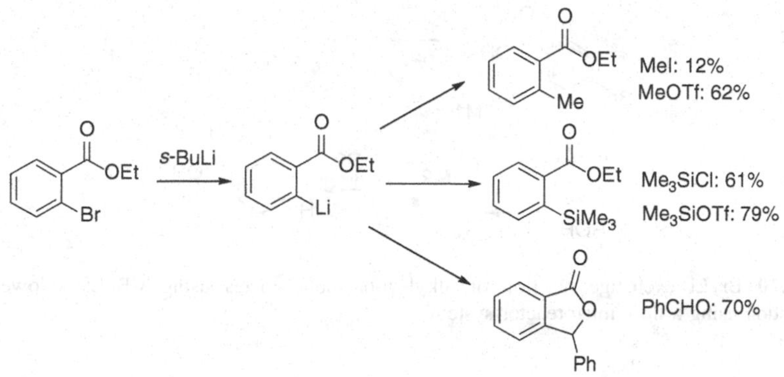

Fig. 4.9 Br–Li exchange reaction of ethyl *o*-bromobenzoates using *s*-BuLi followed by the reaction with an electrophile using a flow microreactor system

reaction that would give no product. This demonstrates that the flow microreactor system enables a chemical transformation that is impossible with a batch reactor.

Aryllithiums bearing alkoxycarbonyl groups generated using the flow microreactor system can be used to react with various electrophiles other than protons. For example, methyl triflate, trimethylsilyl triflate, and benzaldehyde are effective as electrophiles as shown in Fig. 4.9. Notably, the yield greatly differs for different electrophiles. For example, the yield of the methylated product strongly depends on the nature of the methylating agent. For highly reactive methyl triflate, ethyl *o*-methylbenzoate was produced with a yield of 62 %. For less reactive methyl iodide, however, the product was obtained with a yield of 12 %. This is because the reaction with the electrophile competes with the decomposition. When the reaction with the electrophile occurs faster than the decomposition, the desired product is generated in high yields. The yield would be lower when the decomposition occurs faster than the reaction. For the same reason, highly reactive trimethylsilyl triflate allows introduction of the trimethylsilyl group in 79 % yield, whereas less reactive chlorotrimethylsilane allows production of the desired product only in 61 % yield.

4.6 Br–Li Exchange Reactions of Alkyl *p*-Bromobenzoates in a Flow Microreactor [7, 8]

Now, we consider the Br–Li exchange reaction of alkyl *p*-bromobenzoate with *s*-BuLi to generate aryllithium species bearing an alkoxycarbonyl group at *para* position using a flow microreactor system (Fig. 4.10). As can be seen from the temperature–residence time map shown in Fig. 4.11, for the *t*-butyl ester, the protonated product is generated in high yields in a wide range of residence times and temperatures, except that the yield is low in a range of low temperatures and short residence times, where the Br–Li exchange reaction proceeds insufficiently.

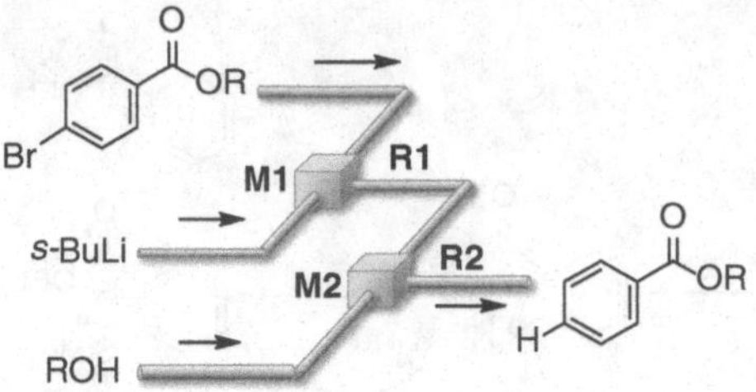

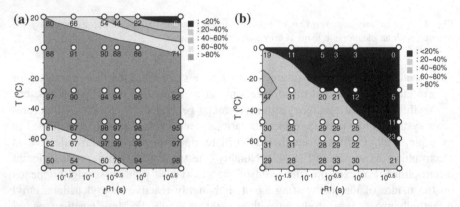

Fig. 4.10 Br–Li exchange reaction of alkyl p-bromobenzoates using s-BuLi followed by protonation using a flow microreactor system

Fig. 4.11 Temperature–residence time map for Br–Li exchange of alkyl p-bromobenzoate followed by protonation, **a** t-butyl p-bromobenzoate and **b** isopropyl p-bromobenzoate

The yield is also low in a range of high temperatures and long residence times, where the decomposition of the aryllithium species occurs.

For the isopropyl ester, the yields were much lower in any range of residence times and temperatures. The decomposition occurs before complete generation of the aryllithium species from the starting material, indicating that the generation and the decomposition cannot be separated by changing the temperature. This also indicates that the generation and reactions of aryllithium species bearing ethoxy-carbonyl and methoxycarbonyl groups in the *para* position are more difficult.

The observations above indicates that the generation of the aryllithium species bearing the alkoxycarbonyl group at the *para* position using Br–Li exchange reaction of alkyl p-bromobenzoates is problematic. In fact, the aryllithium species is known to react with the starting material in batch as shown in Fig. 4.12 [1, 2]. Presumably, in the *ortho* case, the coordination of the alkoxycarbonyl group to Li facilities the generation of the aryllthium species, and therefore, generation is faster the reaction with the starting material. In the *para* case, however, such effect is absent. In addition, the C–Li bond at the *para* position is less sterically hindered than that at the *ortho* position, and therefore, it is more reactive as a nucleophile.

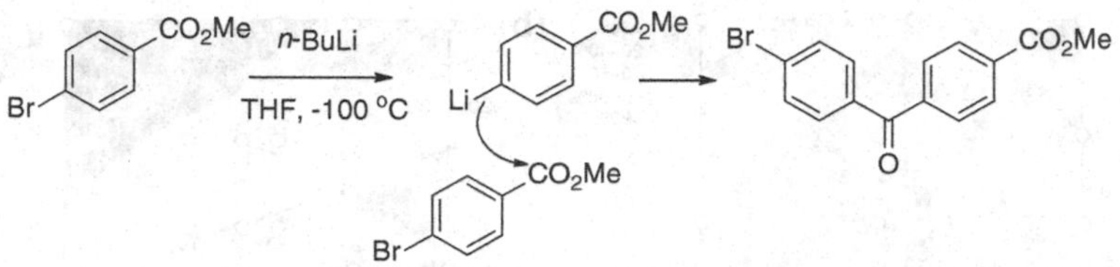

Fig. 4.12 Br–Li exchange reaction of methyl *p*-bromobenzoate with *n*-BuLi in batch

Consequently, the reaction of the aryllithium species and the starting material is inevitable in the *para* case. Thus, we need to change the chemistry for the generation of the aryllithium species.

4.7 I–Li Exchange Reactions of Alkyl *p*-Iodobenzoates in a Flow Microreactor [5]

I–Li exchange reactions are generally much faster than the corresponding Br–Li exchange reactions. The acceleration of the generation process could avoid the undesired reaction with the starting material. Thus, we now consider the I–Li exchange reaction of alkyl *p*-iodobenzoates. However, the use of *s*-BuLi for I–Li exchange may cause a problem, because the resulting *s*-BuI would be a good electrophile to capture the aryllithium species. Therefore, PhLi is used instead of *s*-BuLi, because the resulting PhI is not able to undergo the nucleophilic displacement with the aryllithium species. Thus, I–Li exchange reactions of alkyl *p*-iodobenzoates were conducted using PhLi in the flow microreactor system shown in Fig. 4.13. As shown in the temperature–residence time maps (Fig. 4.14), the protonated products are obtained in good to high yields if we chose an appropriate range for *t*-butyl, isopropyl, ethyl, and methyl esters.

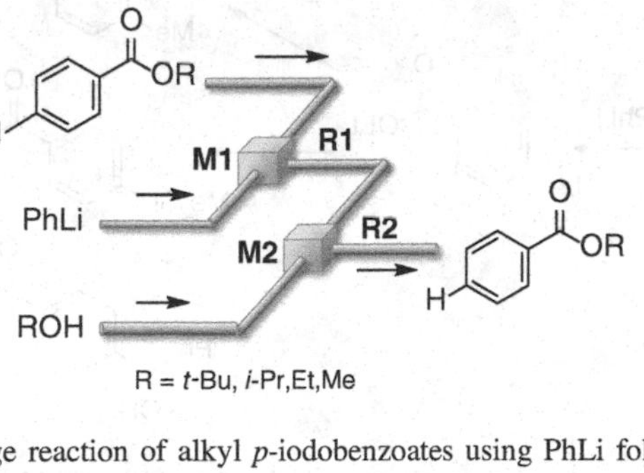

R = *t*-Bu, *i*-Pr, Et, Me

Fig. 4.13 I–Li exchange reaction of alkyl *p*-iodobenzoates using PhLi followed by protonation using a flow microreactor system

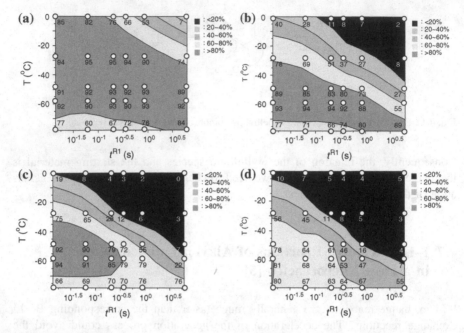

Fig. 4.14 Temperature–residence time map for I–Li exchange of alkyl *p*-iodobenzoates followed by protonation, **a** tert-butyl *p*-iodobenzoate, **b** isopropyl *p*-iodobenzoate, **c** ethyl *p*-iodobenzoate, and **d** methyl *p*-iodobenzoate

Aryllithiums bearing alkoxycarbonyl groups at the *para* position can be used to react with various electrophiles other than protons. As observed for the aryllithium species bearing *o*-alkoxycarbonyl groups, methyl triflate is more effective than methyl iodide. Similarly, trimethylsilyl triflate is more effective than chlorotrimethylsilane (Fig. 4.15).

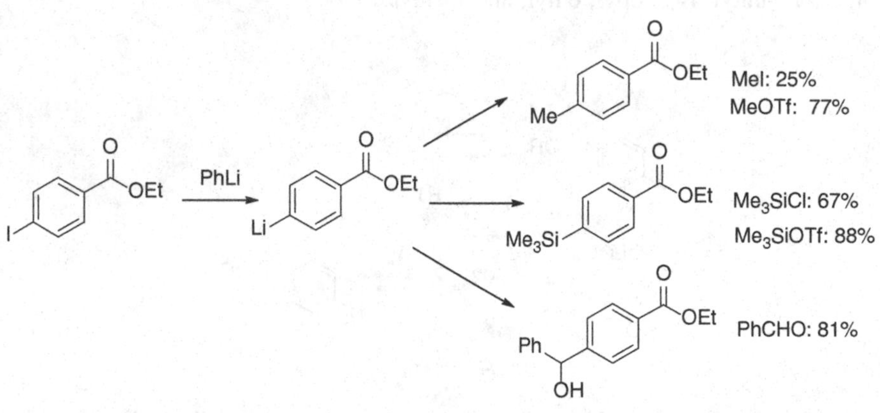

Fig. 4.15 Br–Li exchange reaction of ethyl *p*-iodobenzoates using PhLi followed by the reaction with an electrophile using a flow microreactor system

4.8 Looking Forward

High-resolution reaction time control using a flow microreactor system enables the generation of aryllithium species bearing ester carbonyl groups and allows them to react with electrophiles. This approach leads to protecting-group-free synthesis using flow microreactors, which will be discussed in the next chapter.

References

1. W.E. Parham, Y.A. Sayed, J. Org. Chem. **39**, 2053 (1974)
2. W.E. Parham, L.D. Jones, J. Org. Chem. **41**, 2704 (1976)
3. P. Knochel, W. Dohle, N. Gommermann, F.F. Kneisel, F. Kopp, T. Korn, I. Sapountzis, V.A. Vu, Angew. Chem. Int. Ed. **42**, 4302 (2003)
4. M. Uchiyama, M. Koike, M. Kameda, Y. Kondo, T. Sakamoto, J. Am. Chem. Soc. **118**, 8733 (1996)
5. Y. Kondo, N. Takazawa, C. Yamazaki, T. Sakamoto, J. Org. Chem. **59**, 4717 (1994)
6. F.F. Kneisel, M. Dochnahl, P. Knochel, Angew. Chem. Int. Ed. **43**, 1017 (2004)
7. A. Nagaki, H. Kim, J. Yoshida, Angew. Chem. Int. Ed. **47**, 7833 (2008)
8. A. Nagaki, H. Kim, Y. Moriwaki, C. Matsuo, J. Yoshida, Chem. Eur. J. **16**, 11167 (2010)

Chapter 5
Protecting-Group-Free Synthesis Achieved by High-Resolution Reaction Time Control

Abstract The high-resolution reaction time control using a flow microreactor system allows prompt generation of reactive species bearing highly reactive functional groups in their molecules and enables their use in subsequent reactions before the species decompose. For example, a flow microreactor system allows generation of aryllithium species bearing ketone carbonyl groups and enables them to react with electrophiles before decomposing, although ketones are known to react readily with organolithiums in batch chemistry. Thus, the flow microreactor system enables protecting-group-free synthesis, which involves neither protection nor deprotection, and thus has high atom economy as well as high step economy, enabling synthesis with smaller environmental loads. The present method was successfully applied to formal synthesis of a natural polyphenol, Pauciflorol F.

5.1 Introduction

We have learned that controlling the residence time using a flow microreactor system allows prompt generation of unstable intermediates that can be used in subsequent reactions without decomposing. This concept is applicable to achieving synthesis using no protecting groups. The chapter focuses on shortening the residence time and enabling reactions without protecting the functional groups that would otherwise normally need protection.

5.2 Protecting Group

Producing a complex organic molecule having desired function, such as medical drugs, through a one-step reaction is often difficult and usually needs multistep synthesis involving multiple reactions. In multistep synthesis, the starting materials may contain multiple functional groups, among which one group may be reacted,

© The Author(s) 2015

J. Yoshida, *Basics of Flow Microreactor Synthesis*,
SpringerBriefs in Molecular Science, DOI 10.1007/978-4-431-55513-1_5

but other groups may need to remain unchanged for use in the subsequent reactions. However, such chemoselective reactions are often difficult.

One common approach to solve this problem is to protect functional groups that need to remain unreacted. A functional group to remain intact in one reaction is transformed into another functional group that will be inactive in the reaction (this process is called protection). After the reaction, the functional group is transformed back into the original functional group (this process is called deprotection). Multistep synthesis often involves such protection and deprotection performed repeatedly. The use of a protecting group produces more waste materials because it is discarded upon deprotection, and therefore lowers the atom economy. Atom economy, [1] also called atom efficiency, is indicative of how much of all the atoms in the starting materials constitute the resultant intended products in one chemical process. The concept was proposed by B.M. Trost in 1991 and forms the important basis of environment-friendly chemistry, or green chemistry [2]. Thus, synthesis without using protecting groups and having smaller environmental loads needs to be developed.

Another concept used to determine the efficiency of synthesis is step economy [3]. Synthesizing a complex molecule typically involves multiple steps, and the concept of step economy minimizes the number of steps needed for the synthesis. The use of protecting groups adds two steps, protection and deprotection, and thus lowers the step economy. This generates interest in minimizing the use of protecting groups in synthesis.

A conventional approach toward such protecting-group-free synthesis is to optimize the route of synthesis and minimize the need to use protecting groups [4, 5]. In contrast, a flow microreactor system intends to achieve protecting-group-free synthesis using the same route as for conventional synthesis involving protection and deprotection. High-resolution reaction time control would enable highly chemoselective reactions without using protecting groups. Only one group reacts selectively to achieve an intended transformation, whereas other groups remain unchanged without protection. The principle of such protecting-group-free synthesis will now be described focusing on an organolithium reaction without protecting a ketone carbonyl group.

5.3 Reactions of Organometallic Reagents with Carbonyl Groups [6]

Organometallic reagents including organolithium compounds and Grignard reagents (organomagnesium compounds) can react with various electrophiles and thus can be used widely as strong carbon nucleophiles or carbanion equivalents in organic syntheses. A carbonyl group consists of a carbon–oxygen double bond, and its partially positively polarized carbonyl carbon atom is attacked by a nucleophile such as an organometallic reagent. Therefore, a carbonyl group works as a strong electrophile, and the reaction of an organometallic reagent with a carbonyl compound (in particular a ketone or an aldehyde) serves as one of the most popular reactions in organic synthesis (Fig. 5.1).

We now focus on generating an organolithium species bearing a ketone carbonyl group and reacting it with another electrophile without affecting the carbonyl group. One approach is to protect the carbonyl group by transforming it to another functional group that will be inactive with the aryllithium species. In one common practice, the carbonyl group of a ketone or an aldehyde is protected by transforming it into an acetal, and then, an organolithium reagent is prepared and is reacted with an electrophile such as an aldehyde. The acetal is then transformed back to the carbonyl group through hydrolysis (Fig. 5.2).

The transformation via protecting the carbonyl group is inefficient because it lowers atom economy; ethylene glycol is produced as a by-product upon deprotection. Also, the protection and deprotection increase the number of steps needed for the entire synthesis and lower the atom economy. Intended transformation may be possible without lowering the efficiency, if the generation of the aryllithium species and its reaction with an intended electrophile are both faster than its reaction with the carbonyl group. This can be achieved by promptly preparing the aryllithium species and reacting it with an electrophile before it undergoes decomposition (reaction with the carbonyl group) using a flow microreactor system.

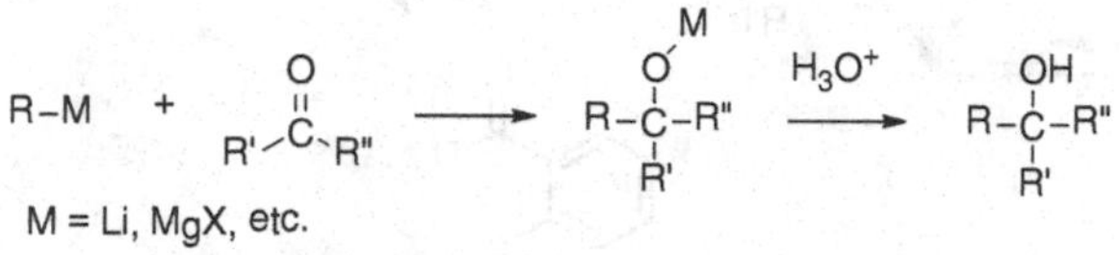

Fig. 5.1 Reactions of organometallic compounds with carbonyl compounds

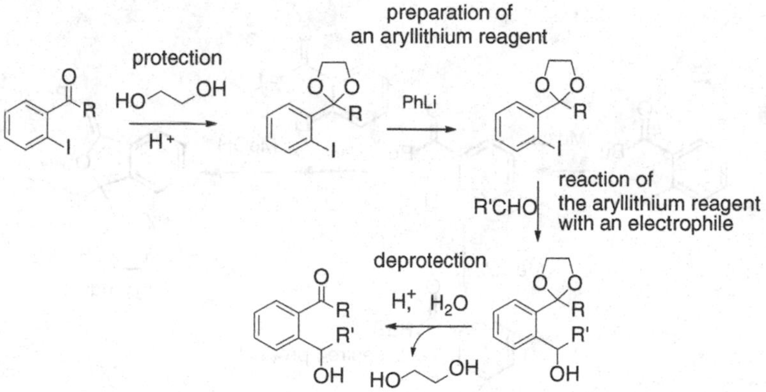

Fig. 5.2 Transformation involving protection, preparation of an aryllithium species, its reaction with an aldehyde, and deprotection

5.4 Generation and Reaction of Aryllithium Species Bearing Ketone Carbonyl Groups

Typically, the carbonyl group of a ketone is highly reactive with a nucleophile. It is thus difficult to generate an aryllithium species containing a ketone carbonyl group and react it with an electrophile without causing reaction of the carbonyl group. This difficultly has been overcome by greatly shortening the residence time (high-resolution reaction time control) using a flow microreactor (Fig. 5.3).

To examine how the residence time affects the results, the following experiment was carried out. Mesityllithium prepared by the reaction of mesityl bromide with *n*-BuLi using a flow microreactor system was reacted with the *o*-iodophenyl butyl

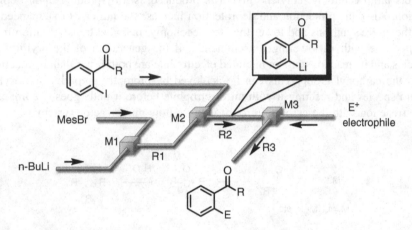

Fig. 5.3 Generation and reactions of aryllithium species bearing ketone carbonyl groups at the *ortho* position

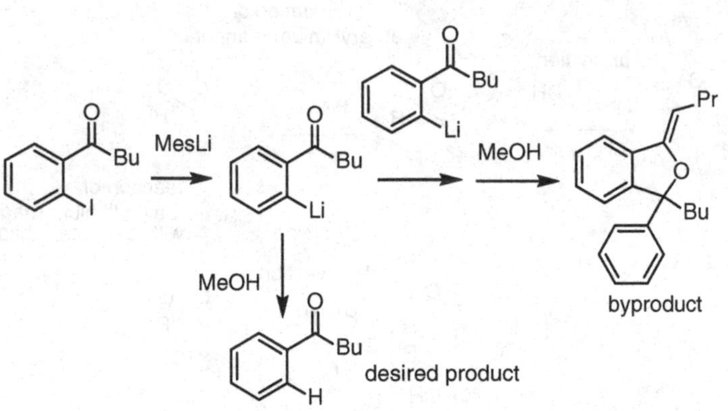

Fig. 5.4 I–Li exchange reaction of *o*-iodophenyl butyl ketone with mesityllithium followed by protonation

ketone to cause an I–Li exchange reaction and to generate the corresponding aryllithium species bearing a ketone carbonyl group. The species was then reacted with methanol (Fig. 5.4). The yield of the protonated product increases as the residence time is shorter (Fig. 5.5). The yield of the by-product (resulting from dimerization of the aryllithium species) decreases accordingly. Only shortening the residence time to as short as 3 ms using a built-in flow microreactor system consisting of two micromixers and one flow microreactor (Fig. 5.6) has produced the protonated product in a yield of 90 % or more.

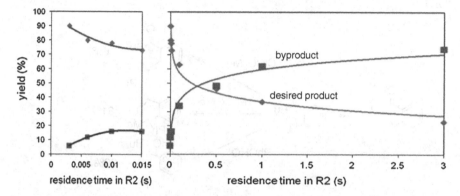

Fig. 5.5 The effect of the residence time in **R2** on the yields of the desired protonated product and by-product in I–Li exchange reaction of *o*-iodophenyl butyl ketone followed by protonation

Fig. 5.6 A built-in-type flow microreactor system consisting of two micromixers (**M2** and **M3**) and a flow microreactor (**R2**)

Various aryllithiums bearing ketone carbonyl groups were generated under the optimum conditions and were reacted with various electrophiles. The results are shown in the Fig. 5.7. Although their reactions with the electrophiles need to be faster than their decomposition (dimerization between aryllithium species), a variety of electrophiles including alkyl chloroformate, aldehydes, highly reactive ketones, and isocyanates can be used. In addition, phenyllithium bearing ketone carbonyl groups at the *para* position can be generated and used for the reaction with electrophiles. It is amazing that generation and reaction of a phenyllithium containing methyl ketones at the *para* position was successfully accomplished,

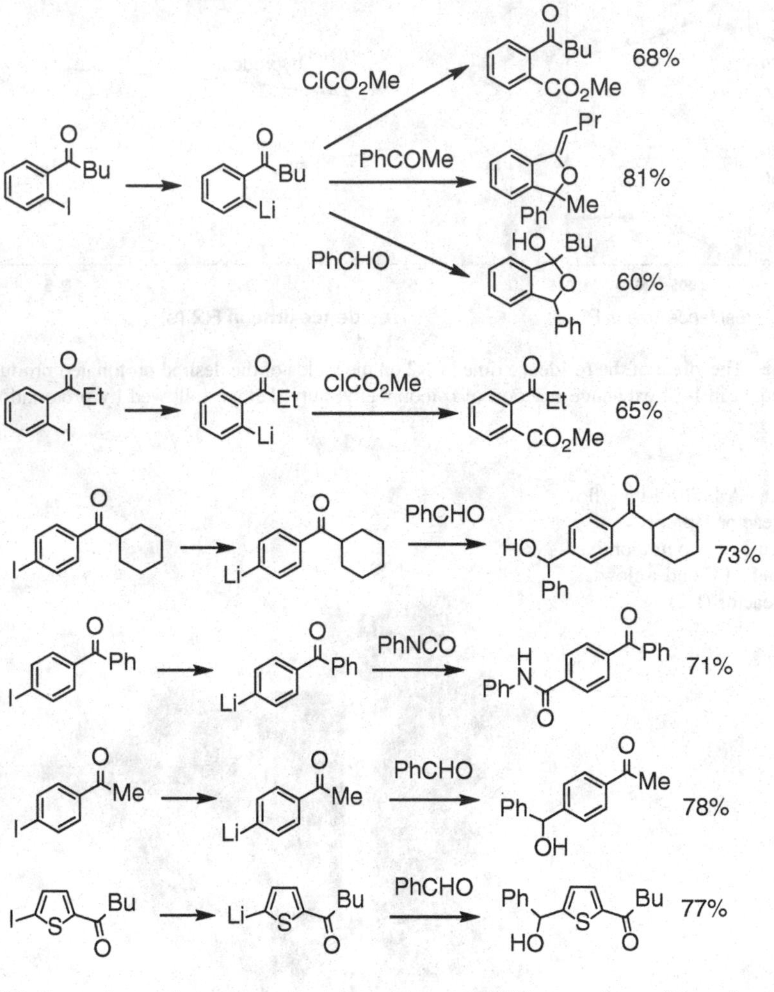

Fig. 5.7 Generation and reactions of aryl- and heteroaryllithiums using a flow microreactor system

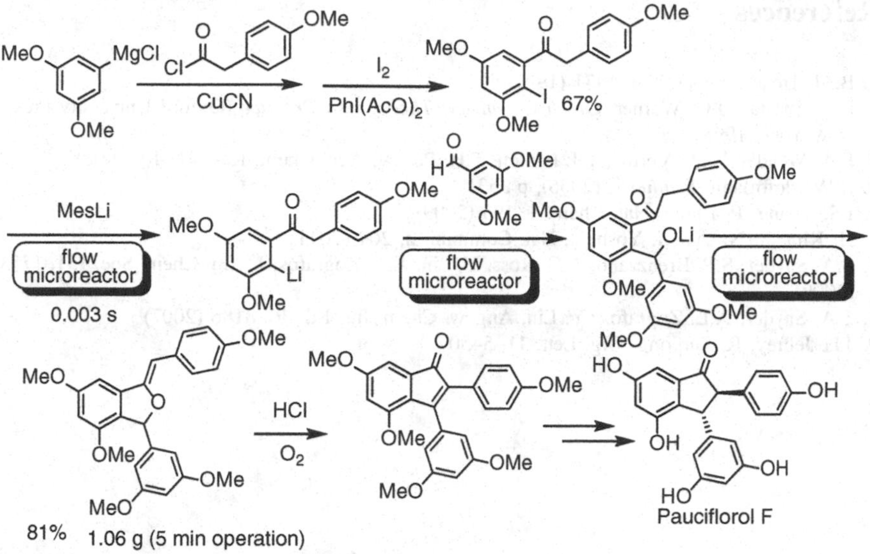

Fig. 5.8 Synthesis of Pauciflorol F

although the residence time needs to be decreased further to 1.5 ms. Moreover, heteroaryllithiums bearing ketone carbonyl groups were also successfully generated and reacted with the electrophiles.

The present protecting-group-free method has further enabled formal synthesis of Pauciflorol F,[1] which is one of the natural polyphenols (Fig. 5.8). The flow microreactor system eliminates the steps of protecting and deprotecting the ketone carbonyl group and has allowed synthesis with less steps. The starting material for I–Li exchange was prepared from commercially available 3,5-dimethoxyphenyl-magnesium chloride in two steps. The generation of the corresponding aryllithium species was conducted using a flow microreactor system consisting of the built-in-type integrated device. The I–Li exchange reaction took place selectively without affecting the ketone carbonyl group. Also, the acidic methylene group *alpha* to the carbonyl was not affected appreciably. A product derived from enolized interme-diate was not observed. The reaction with 3,5-dimethoxybenzaldehyde afforded the desired product in 81 % isolated yield, via dehydration upon the acidic workup. Notably, 1.06 g of the product was obtained by 5-min operation. Treatment with HCl/*i*-PrOH in the presence of O_2 in a batch reactor gave the cyclopentenone derivative in 75 % yield, which is known to be converted to Pauciflorol F via hydrogenation and epimerization followed by deprotection of methyl ethers.

---

[1]Total synthesis of Pauciflorol F using the flask method [7–9].

References

1. B.M. Trost, Science **254**, 1471 (1991)
2. P.T. Anastas, J.C. Warner, *Green Chemistry: Theory and Practice* (Oxford University Press, New York, 1998)
3. P.A. Wender, V.A. Verma, T.J. Paxton, T.H. Pillow, Acc. Chem. Res. **41**, 40 (2008)
4. R.W. Hoffmann, *Synthesis*, (2006), p 3531
5. I.S. Young, P. Baran, Nat. Chem. **1**, 193 (2009)
6. H. Kim, A. Nagaki, J. Yoshida, Nat. Commun. **2**, 264 (2011)
7. S.A. Snyder, S.P. Breazzano, A.G. Ross, Y. Lin, A.L. Zografos, J. Am. Chem. Soc. **131**, 1753 (2009)
8. S.A. Snyder, A.L. Zografox, Y. Lin, Angew. Chem. Int. Ed. **46**, 8186 (2007)
9. J.L. Jeffrey, R. Sarpong, Org. Lett. **11**, 5450

Chapter 6
Controlling Isomerization by High-Resolution Reaction Time Control

Abstract When a reactive intermediate can be isomerized and the isomerized intermediate can further react to give an isomer of an intended product, setting a shorter residence time will suppress such isomerization and allow highly selective synthesis of an intended product. The rate of isomerization is also temperature dependent, and thus, controlling not only the residence time but also the temperature is important. In some cases, the isomerization may proceed completely to give a more stable isomer of the reactive intermediate. In such cases, high-resolution reaction time control allows selective synthesis of either a product derived from the initial intermediate or a product derived from the isomerized intermediate at will. When isomerization of an intermediate gives an equilibrium mixture, setting a shorter residence time will suppress such isomerization and allow highly selective synthesis of a product derived from the intermediate that has yet to be isomerized. This concept basically can be applied to positional isomers and stereoisomers. In particular, this concept is useful in asymmetric synthesis in which epimerization of an optically active intermediate can be suppressed to allow synthesis of a product with high enantiomeric purity.

6.1 Introduction

We have learned in Chap. 4 that high-resolution reaction time control using flow microreactors allows short-lived reactive species to be usable before they decompose, and in Chap. 5 that this concept is applicable to protecting-group-free synthesis. Short-lived reactive species can undergo not only decomposition but also isomerization. The isomerization can be controlled by using a flow microreactor system and setting appropriate residence times precisely. This chapter describes controlling isomerization through high-resolution reaction time control, taking three types of isomerization of organolithium species as examples.

© The Author(s) 2015
J. Yoshida, *Basics of Flow Microreactor Synthesis*,
SpringerBriefs in Molecular Science, DOI 10.1007/978-4-431-55513-1_6

6.2 Controlling Positional Isomers [1]

Nitro groups are powerful electron-withdrawing groups and are important functional groups in organic synthesis. For example, they can be easily reduced to amino groups, which often exist in biologically active compounds and functional materials. One common approach to introduce a nitrogen functional group to an aromatic ring is first introducing a nitro group by nitration and then converting the nitro aromatics to another nitrogen-containing aromatic compound. However, nitro groups react easily with organolithiums, and it has thus been considered difficult to generate and use an organolithiums bearing a nitro group. Now, high-resolution reaction time control using a flow microreactor system allows generation of an aryllithium species bearing a nitro group and enables its use in subsequent reactions with electrophiles.

Aryl bromide **A** shown in Fig. 6.1 is reacted with phenyllithium to produce the corresponding aryllithium species, which is then reacted with an aldehyde. Within a residence time of about 0.06 s, aryl bromide **A** was consumed completely to give the corresponding product **B** with a high yield. However, a longer residence time will cause the lithium to migrate to a position where both the nitro group and the methoxy group can coordinate producing a more stable aryllithium species. This isomerization is almost complete within a residence time of 63 s. An aldehyde is then added to capture the isomerized aryllithium species to give product **C**.

As described above, high-resolution reaction time control enables switching product selectivity at will. A product derived from a reactive species that has yet to be isomerized can be obtained by setting a shorter reaction time, or a product derived from an isomerized reactive species can be obtained by setting a longer reaction time.

6.3 Controlling Diastereomers [2, 3]

For some oxiranyllithium species, which we have discussed in Chap. 2, two stereoisomers can exist. In such cases, an oxiranyllithium species has another side reaction that needs attention: Isomerization caused by inversion of configuration at its carbon atom bound to lithium in addition to its decomposition involving ring opening. To synthesize a substituted epoxide stereoselectively, we need to suppress such isomerization and allow only one isomer to selectively react with an electrophile. For this purpose, high-resolution reaction time control using a flow microreactor system is effective.

Such isomerization can be observed when a trisubstituted oxirane shown in Fig. 6.2 is used. In more detail, oxiranyllithium species c–**E**, which results from deprotonation of (2R*,3S*)-2-methyl-2,3-diphenyloxirane (c–**D**), reacts with methyl iodide to give tetrasubstituted oxirane c–**F**. When, however, the species c–**E** isomerizes to t–**E**, t–**E** reacts with methyl iodide to give the corresponding

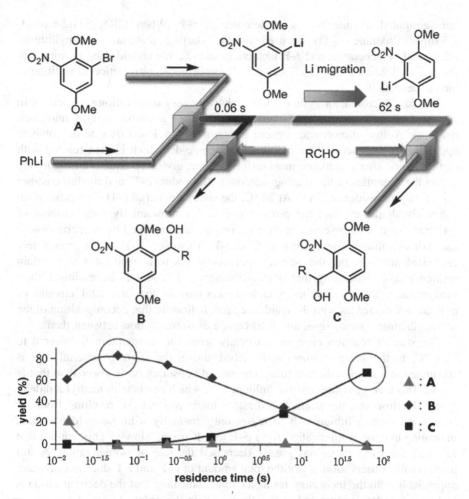

Fig. 6.1 Control of isomerization of the aryllithium species by the residence time

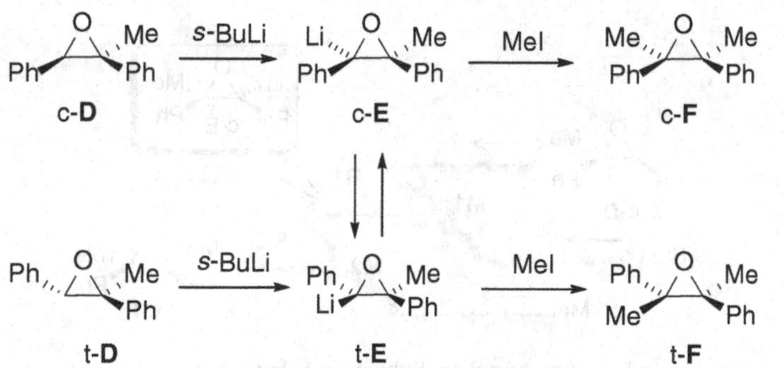

Fig. 6.2 Generation and reaction of the oxiranyllithiums with isomerization

tetrasubstituted oxirane *t*–**F**, a diastreomer of *c*–**F**. When (2R*,3S*)-2-methyl-2,3-diphenyloxirane (*t*–**D**) is used as the starting material, oxiranyllithium *t*–**E** would be generated, and *t*–**E** isomerizes to *c*–**E**. We should keep in mind that the oxiranyllithium species *c*–**E** and *t*–**E** also undergo decomposition in addition to the isomerization.

The flow microreactor system allows relatively easy observation of reactions in such complicated circumstances and thus advances a deeper insight into such reactions. A flow microreactor system shown in Fig. 6.3 was used to deprotonate epoxide *c*–**D** to generate oxiranyllithium species *c*–**E**, which is then reacted with methyl iodide after a predetermined residence time, giving epoxide *c*–**F**. Figure 6.4 shows the quantities of the starting material *c*–**D**, product *c*–**F**, and its diastereomer *t*–**F** at varying residence times. At 24 °C, the starting material *c*–**D** disappears in an extremely short time and the product *c*–**F** and a substantially large amount of *t*–**F** are formed. This reveals that the reactions are accompanied by isomerization of the oxiranylltihium intermediate *c*–**E** to *t*–**E**. However, *c*–**E** is not completely converted into *t*–**F**, but the isomerization converges to a fixed ratio at a certain residence time, indicating that oxiranyllithiums *c*–**E** and *t*–**E** have similar thermodynamic stability. Another notable observation is that the total amount of *c*–**F** and *t*–**F** decreases with the residence time, indicating that decomposition of the oxiranyllithium species (*c*–**E** and *t*–**E**) competes isomerization between them.

The state of reaction changes drastically when the temperature is lowered to −28 °C. In the same manner as described above, the starting material *c*–**D** is consumed in a short residence time. The rate of consumption is lower than that at 24 °C. This then generates oxiranyllithium *c*–**E**, which reacts with methyl iodide to give *c*–**F**. However, the amount of diastereomeric product *t*–**F** resulting from the isomerized oxiranyllithium *t*–**E** increases only gradually with the residence time, indicating that the isomerization from *c*–**E** to *t*–**E** occurs slowly. This implies that lowering the temperature has greatly decreased the rate of isomerization. A still more notable observation is that the total amount of *c*–**F** and *t*–**F** does not decrease appreciably with the increasing residence time, indicating that the decomposition is almost completely suppressed in this timescale (in the order of seconds).

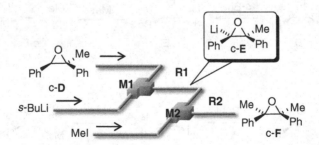

Fig. 6.3 Generation and reaction of oxiranyllithium in a flow microreactor system

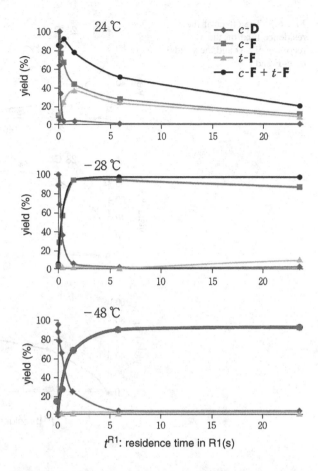

Fig. 6.4 The effect of the residence time on the recovery of *c*–**D** and the yields of *c*–**F** and *t*–**F**

When the temperature is lowered further to −48 °C, the isomerization can be suppressed almost completely. As a result of this, the starting material *c*–**D** can be deprotonated to generate oxiranyllithium species *c*–**E**, which can then react with methyl iodide to selectively give product *c*–**F**, without any decomposition and isomerization of *c*–**E**.

Figure 6.5 shows the results of similar experiments using (2R*,3R*)-2-methyl-2,3-diphenyloxirane (*t*–**D**). Although the starting material *t*–**D** is deprotonated more slowly than its diastereomer *c*–**D**, the experimental results show the same overall tendency as described above.

Making a temperature–residence time map for such reactions is useful for controlling the reactions at optimum temperatures and with optimum residence times enabling selective synthesis of desired products with high yields.

Fig. 6.5 The effect of the residence time on the recovery of *t*–**D** and the yields of *c*–**F** and *t*–**F**

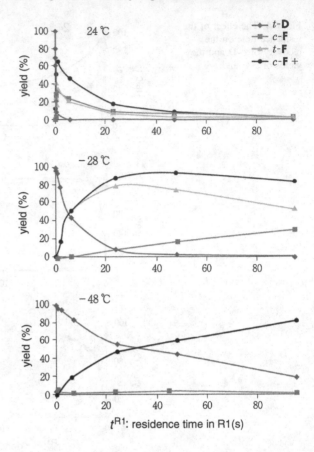

6.4 Controlling Stereoisomers (Enantiomers) [4]

The concept of high-resolution reaction time control using a flow microreactor system can also be effectively applied to asymmetric synthesis. In some cases, enantiomerically enriched intermediates can easily be isomerized (racemized) reducing the enantioselectivity of the reaction.[1] Batch reaction would often give products with low enantiomeric purity because of such isomerization. Flow microreactor system, however, could allow the formation of products with high enantiomeric purity by first generating the intermediates and then promptly using it in subsequent reactions before isomerization. An example of such synthesis will now be described.

---

[1]For the stereochemical control of chiral organolithium species: Lee et al. [5] and references cited therein.

An enyne compound (**G**) containing a directing group CbO (CONPr-i_2) that can coordinate lithium is reacted with n-BuLi in the presence of sparteine (**L***), which is an amine commercially available in enantiomerically enriched form. The butyl group of n-BuLi is added to the carbon–carbon double bond of **G** to generate an enantiomerically enriched organolithium intermediate (**H**) (Fig. 6.6). Intermediate **H** epimerizes easily into **H'**. When intermediate **H** is reacted with an electrophile, the acetylene terminus forms a bond, thus producing an enantiomerically enriched allene (**I**). The Hoffmann test [6–8] shows that intermediate **H** reacts with an electrophile at a rate higher than the rate of its isomerization into **H'**, and also shows that the enantiomeric purity of the resultant allene is determined by the ratio of **H** and **H'**.

This reaction was conducted using a flow microreactor system shown in Fig. 6.7 (quenched with methanol, where E = H). Figure 6.8 shows the yields and enantiomeric composition of the product [9] with varying the residence time in reactor **R2**. At low temperatures and with short residence times, the reaction proceeds insufficiently, resulting in a low yield of the product. At high temperatures and with long residence times, the product is obtained with a good yield, but the enantiomeric composition (ec) is low, seemingly because the isomerization of intermediate **H** into

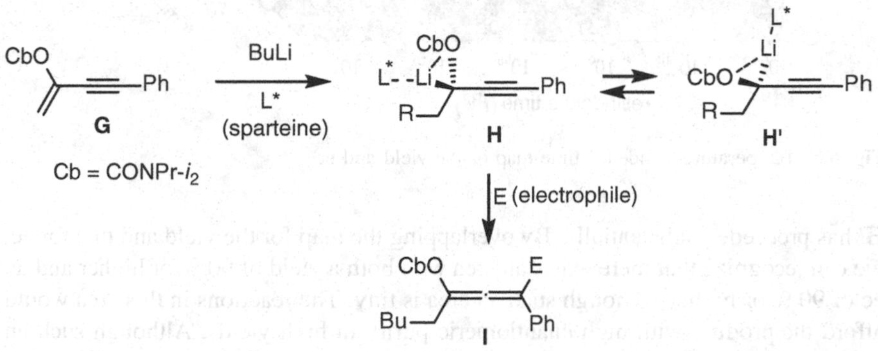

Fig. 6.6 Synthesis of enantiomerically enriched allenes by carbolithiation of the eneyne followed by the reaction with electrophiles

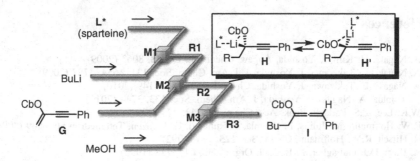

Fig. 6.7 Synthesis of enantiomerically enriched allenes using a flow microreactor system

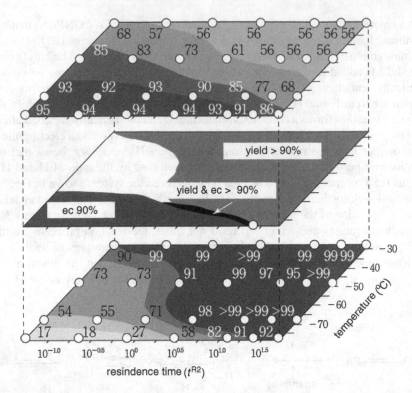

Fig. 6.8 Temperature–residence time map of the yield and ec

H′ has proceeded substantially. By overlapping the map for the yield and that for ec, we can recognize that there exists an area with both a yield of 90 % or higher and an ec of 90 % or higher, although such an area is tiny. The reactions in this area would afford the product with high enantiomeric purity in high yields. Although such an area may not always be found, this example demonstrates that a flow microreactor system may enable rational development of asymmetric reactions.

References

1. A. Nagaki, H. Kim, J. Yoshida, Angew. Chem. Int. Ed. **48**, 8063 (2009)
2. A. Nagaki, E. Takizawa, J. Yoshida, J, J. Am. Chem. Soc. **131**, 1654, 3787 (2009)
3. A. Nagaki, E. Takizawa. J. Yoshida, Chem. Eur. J. **16**, 14149 (2010)
4. Y. Tomida, A. Nagaki, J. Yoshida, J. Am. Chem. Soc. **133**, 3744 (2011)
5. W.K. Lee, Y.S. Park, P. Beak, Acc. Chem. Res. **42**, 224 (2009)
6. R.W. Hoffmann, M. Julius, F. Chemla, T. Ruhland, G. Frenzen, Tetrahedron **50**, 6049 (1994)
7. R. Hirsch, R.W. Hoffmann, Chem. Ber. **125**, 975 (1992)
8. A. Basu, D.J. Gallagher, P. Beak, J. Org. Chem. **61**, 5718 (1996)
9. H.B. Kagan, Recl. Travl. Chim. Pays-Bas **114**, 203 (1995)

Chapter 7
Controlling Competitive Consecutive Reactions Using Micromixing

Abstract When the reaction is faster than mixing in competitive consecutive reactions, the selectivity of the product may not be determined by the kinetics but by the way of mixing. This is called disguised chemical selectivity. Mixing using a batch macroreactor is typically slow, and disguised chemical selectivity is often observed for fast reactions. Mixing using a flow microreactor system incorporating a micromixer is faster and thus can effectively solve this problem, allowing the reaction to stop at the first stage with high selectivity in many cases. Micromixing is effective for improving the product selectivity of Friedel–Crafts-type reactions and other types of reactions.

7.1 Introduction

In Chaps. 4–6, we have learned examples of the residence time control allowing unstable intermediates to form in short times and react with other compounds before decomposing. In other words, the product selectivity depends on the residence time and the temperature, although fast micromixing is essential because the residence time control requires mixing to be complete in a shorter time than the residence time as we have learned in Chap. 3. This chapter explores another aspect of the chemistry of flow microreactor synthesis, where extremely fast mixing using a micromixer directly affects the product selectivity. In this case, the product selectivity is not determined by the kinetics but by the mixing speed, if the reaction is very fast. In other words, the product selectivity strongly depends on the speed of mixing. We now focus on fast competitive consecutive reactions, in which a product molecule from the first reaction further reacts with another molecule of the starting material to give a second product.

© The Author(s) 2015

J. Yoshida, *Basics of Flow Microreactor Synthesis*,
SpringerBriefs in Molecular Science, DOI 10.1007/978-4-431-55513-1_7

7.2 Factors Determining Selectivity in Chemical Reactions

One major goal of synthetic chemistry is to selectively synthesize desired compounds, but it is often difficult to selectively obtain only the desired products without producing any undesired products. Selectivity in chemical reactions can mainly be categorized into two cases: selectivity for starting materials and selectivity for products (product selectivity). The selectivity for starting materials refers to the favored reaction of some starting materials over others in mixing multiple starting materials. To synthesize a specific compound, usually only needed starting materials are used and we do not mix other materials. Thus, the selectivity for starting materials can rarely be a problem in chemical synthesis. The selectivity for products refers to the favored generation of products over other products from one starting material. The product selectivity can be a problem in chemical synthesis. The product selectivity can range from positional selectivity and functional group selectivity (chemoselectivity) to enantioselectivity. Controlling such selectivity through the residence time control has been discussed in Chaps. 4–6. This chapter describes controlling the product selectivity in a slightly different point of view.

The selectivity for products is often determined by the rate of reaction to give those products. The reaction rate typically depends on the activation energy and the concentration of compounds involved in the reaction. To improve the product selectivity, an auxiliary group such as protecting group or directing group is often used to change the activation energy. However, the selectivity may not always be achieved in accordance with the reaction kinetics. The kinetics holds only when the concentration of chemical species involved in the reaction is uniform and the temperature is also uniform. When, for example, a reaction is conducted by mixing two reaction components, but they react extremely rapidly before the mixing is complete, the selectivity would not be achieved in accordance with the kinetics. This chapter describes such reactions.

7.3 Competitive Consecutive Reactions

Let us consider a reaction in which compounds **A** and **B** react with each other to give product **P1**, which then further reacts with compound **B** to give second product **P2** (Fig. 7.1). Therefore, this is a consecutive reaction. The selectivity in affording either **P1** or **P2** is the product selectivity as viewed from **A**, whereas the selectivity

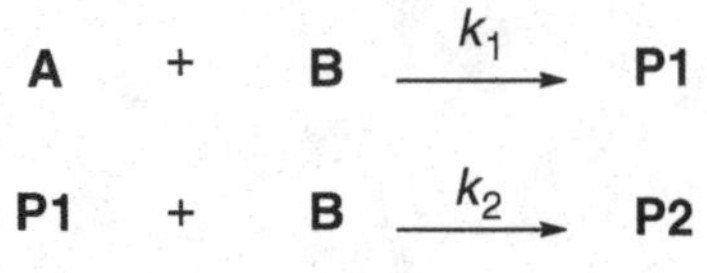

$$\text{A} \quad + \quad \text{B} \xrightarrow{\ k_1\ } \text{P1}$$

$$\text{P1} \quad + \quad \text{B} \xrightarrow{\ k_2\ } \text{P2}$$

Fig. 7.1 General scheme of competitive consecutive reactions

of either **A** or **P1** reacting with **B** is the selectivity for starting materials (vide supra). The reaction of **A** with **B** and the reaction of **P1** with **B** compete with each other. In this context, this is also a competitive reaction. In other words, the first-stage reaction competes with the second-stage reaction in the consecutive reactions. Such reactions are called the competitive consecutive reactions.

In the competitive consecutive reactions, selectively producing **P2** is easy. If the second-stage reaction is slower than the first-stage reaction, an excessive amount of **B** can be used to extend the reaction for a sufficiently long time. When the second-stage reaction is faster, selectively producing **P2** thus has no problem, because **P1** reacts with **B** immediately after it is generated. Selectively producing the first product **P1**, however, is not always easy. We now consider this case. The use of an excessive amount of **B** would allow easy production of the second product **P2**, as mentioned above. The compounds **A** and **B** thus need to react with each other at the molar ratio of 1:1. Alternatively, an excessive amount of compound **A** may be used if such use is acceptable. In this case, the excess **A** will remain unchanged. If the second-stage reaction is fast, the second product **P2** would inevitably form as a by-product. If, however, the first-stage reaction is sufficiently faster than the second-stage reaction, the desired **P1** would be obtained selectively. This is the selectivity based on the kinetics. However, if the reaction of **A** and **B** is faster than their mixing, the reaction occurs before the mixing gives a homogeneous solution. The selectivity in this case will now be described.

7.4 Disguised Chemical Selectivity [1, 2]

Adding the solution of **B** into the solution of **A** would not immediately cause mixing between these compounds. When a drop of the solution of **B** falls into the solution of **A**, the initial drop is broken by stirring, but **B** still stays as small masses. The state (a) illustrated in Fig. 7.2 shows one of them. When the reaction is faster

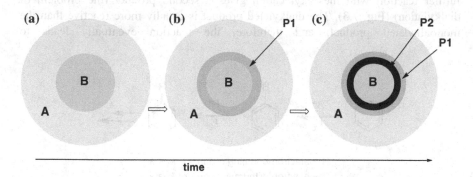

Fig. 7.2 A scheme for a competitive consecutive reaction of **A** with **B** to give products **P1** and **P2** in the case that the reaction is faster than mixing

than the mixing and the reaction proceeds in this state, the reaction will occur at the interface between the solutions of **A** and **B**. The first product **P1** from the first-stage reaction forms at the interface between the solutions of **A** and **B** as illustrated in (b). This means that compound **B** cannot react with **A** because they are separated by the product **P1**. Compound **B** meets the first product **P1** to produce the second product **P2** as illustrated in (c) even when the first-stage reaction is faster than the second-stage reaction. The resultant selectivity would differ from the selectivity predicted based on the kinetics. This phenomenon is called disguised chemical selectivity.

In batch macroreactors, the reactions would not show the predicted selectivity based on the kinetics but often have lower selectivity due to this phenomenon. The problem of disguised chemical selectivity can be solved only either by decreasing the reaction rate or by increasing the mixing speed. Traditional approaches to this problem focus on lowering the temperature, or lowering the concentration to decrease the reaction rate. However, such approaches are not preferred because the time efficiency or productivity becomes lower. Increasing the mixing speed is a better approach. You may remember that micromixers described in Chap. 3 can be used to increase the mixing speed.

7.5 Controlling Selectivity of Friedel–Crafts-Type Reactions Using Micromixing [3, 4]

The Friedel–Craftsl-type reaction of an aromatic compound will now be discussed as an example reaction in which the problem of selectivity in competitive consecutive reactions has been solved by micromixing. The Friedel–Crafts-type reaction is a reaction of an aromatic compound such as benzene with an organic cationic species. In the Friedel–Crafts alkylation reaction (the reaction with alkyl cations), a product is more reactive than the starting aromatic compound, because an alkyl group is usually electron-donating and activates the aromatic ring toward electrophilic reactions. Therefore, the reaction cannot be stopped at that stage, and a further reaction with the alkyl cation gives a second product (the problem of dialkylation) (Fig. 7.3). The dialkylated product is usually more reactive than the monoalkylated product, and therefore, the reaction eventually leads to

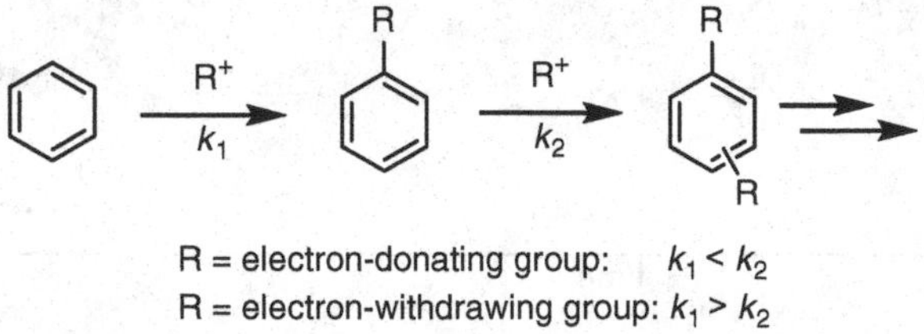

R = electron-donating group: $k_1 < k_2$
R = electron-withdrawing group: $k_1 > k_2$

Fig. 7.3 Product selectivity of Friedel–Crafts reactions

polyalkylation (the problem of polyalkylation). Stopping the reaction at the stage of monoalkylation usually requires an excessively large amount of an aromatic compound. In some practical examples, the reaction may be controlled by using an aromatic compound in an amount of a solvent. In contrast, when an acyl cation is reacted, the reaction can often be stopped at the first stage because the acyl group is electron-withdrawing, and thus, the monoacylation product would be less reactive than the starting aromatic compound.

Typically, organic cationic species, such as alkyl cations or acyl cations, are generated from its precursor such as an organic halide by the treatment with a Lewis acid. This reaction proceeds in equilibrium, and the equilibrium is often predominantly shifted to the precursor because an organic cation is usually very unstable. Therefore, the solution in equilibrium contains an organic cation at an extremely low concentration. The organic cation consumed as the reaction proceeds is simply supplied based on the equilibrium. To allow an aromatic compound and an organic cation to react at the ratio of 1:1, their mixing needs to be at 1:1. However, mixing the precursors of the organic cation with the aromatic compound at 1:1 would not be equivalent to mixing the organic cation and the aromatic compound at 1:1.

To accurately mix the aromatic compound and the organic cation at 1:1 molar ratio and allow them to react, the organic cation need to be stored in a solution at a relatively high concentration and then mixed with a solution of the aromatic compound. However, the organic cation are usually unstable, except some stable cations such as trityl cations that are stabilized by the neighboring aromatic rings. Usually, the unstable organic cations cannot be stored in a solution without using a special medium, such as a superacid. Organic cations are thus normally generated in situ through an equilibrium reaction in the presence of a nucleophile to be reacted (vide supra).

However, the use of low-temperature electrochemical oxidation allows generation and storage of organic cations such as N-acyliminium ions in a normal reaction medium such as dichloromethane in the absence of a nucleophile. This method is called the cation pool method (Fig. 7.4) [5–7].

The organic cations generated and accumulated by the cation pool method can be observed by nuclear magnetic resonance (NMR) spectroscopy or infrared spectroscopy (IR). A solution can contain organic cations at concentrations of 0.1–0.05 M. An N-acyliminium ion generated by the cation pool method can be used for the Friedel–Crafts alkylation reaction of aromatic compounds.

Before using micromixing, conventional stirring in a flask was first examined (Fig. 7.5). 1,3,5-Trimethylbenzene was reacted with 1 equivalent of the

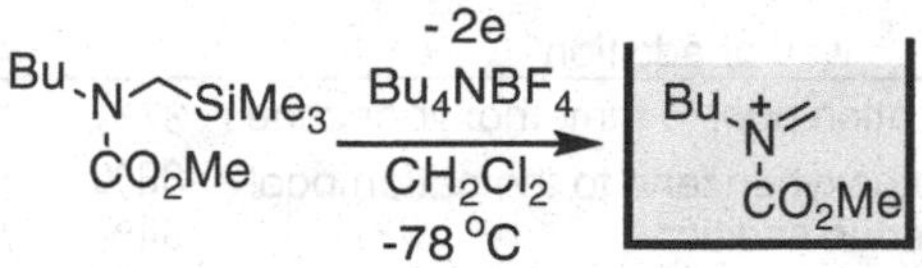

Fig. 7.4 Generation and accumulation of the N-acyliminium ion by the cation pool method

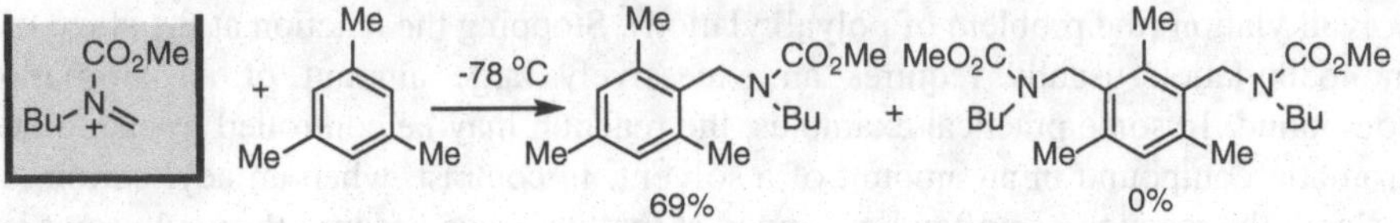

Fig. 7.5 Reaction of the *N*-acyliminium ion with 1,3,5-trimethylbenzene in batch

N-acyliminium ion prepared by the cation pool method. The molar ratio of the cations and the aromatic compound was set to 1:1. This gave a product reacted at 1:1 with a yield of 69 % and gave no product reacted at 1:2.

An *N*-acylaminomethyl group added to the benzene ring through this reaction is typically an electron-donating group. In the Friedel–Crafts alkylation reaction in which an electron-donating group is added, the reaction should not be stopped at the first stage, allowing dialkylation and polyalkylation to occur easily. However, this reaction has stopped after the monoalkylation. This is presumably because a proton removed from the benzene ring in the reaction protonates the *N*-acylaminomethyl group, which then becomes an electron-withdrawing group. The molecular orbital theory reveals that the reaction would produce the compound in which carbonyl oxygen of the *N*-acylaminomethyl is protonated. In this case, the second-stage reaction is slower than the first-stage reaction, and thus, the reaction seems to stop after the monoalkylation.

The same reaction was conducted using 1,3,5-trimethoxybenzene, which is much more nucleophilic than 1,3,5-trimethylbenzene (Fig. 7.6). In this case, even when the two species were allowed to react at 1:1 molar ratio, a large amount of dialkylated product was formed. Changing the manner of mixing, or reversing the order of adding one solution to the other, would have the same result. The *N*-acylaminomethyl group in the product was seemingly protonated also in this case, suggesting that the reaction is faster than the mixing because 1,3,5-trimethoxybenzene is more nucleophilic than 1,3,5-trimethylbenzene and that the faster reaction causes the problem of disguised chemical selectivity.

To mix the two reaction components much faster, a flow system incorporating a micromixer was used to conduct the reaction (Fig. 7.7). With a conventional simple T-tube (ϕ = 500 μm) used as a mixer, the achieved selectivity was not much

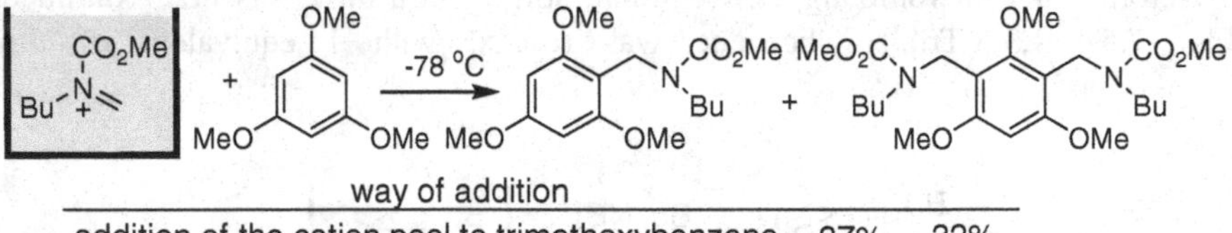

way of addition		
addition of the cation pool to trimethoxybenzene	37%	32%
addition of trimethoxybenzene to the cation pool	33%	33%
simultaneous addition	34%	30%

Fig. 7.6 Reaction of the *N*-acyliminium ion with 1,3,5-trimethoxybenzene in batch

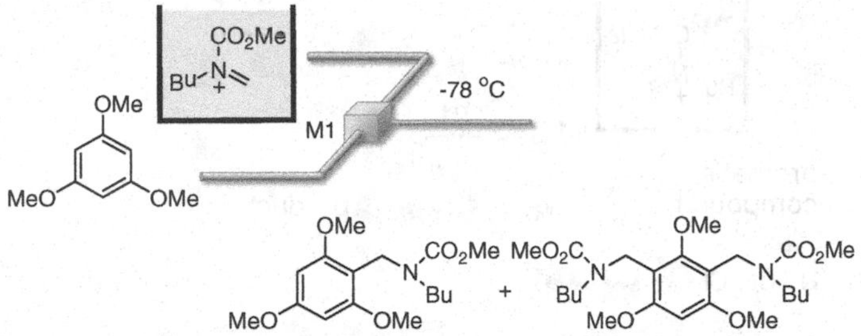

mixer			
conventional T-tube (φ = 500 μm)	36%	31%	(54 : 46)
YM-1 micromixer	50%	14%	(78 : 22)
IMM micromixer (channel width = 25 μm)	34%	30%	(96 : 4)

Fig. 7.7 Reaction of the N-acyliminium ion with 1,3,5-trimethoxybenzene using a micromixer

different from the selectivity achieved with a flask. With a YM-1 micromixer, the selectivity of the monoalkylation product was improved. With an IMM micromixer (channel width: 25 μm), the selectivity of the monoalkylation product was markedly high. The increased mixing speed seems responsible for solving the problem of disguised chemical selectivity and achieving the selectivity close to the targeted selectivity based on the kinetics. The importance of the mixing speed is also demonstrated by experiments conducted with the varying flow rates. Even with the IMM micromixer, a small flow rate results in a low selectivity. The IMM micromixer is known to have high mixing performance at flow rates not less than 5 mL min^{-1}. In this reaction as well, the selectivity is sufficiently high at a flow rate of 5 mL min^{-1}. As demonstrated by the large impact of the flow rates, the mixing speed is important in improving the product selectivity.

The experiment below will now demonstrate that the first-stage reaction is really faster than the second-stage reaction. In the experiment, a flow microreactor system incorporating the IMM micromixer (**M1**) was used for the reaction of 1,3,5-trimethoxybenzene and the N-acyliminium ion (Fig. 7.8). The residence time was set to 0.03 s for a reactor **R1**, affording the monoalkylation product in 78 % yield after workup. The monoalkylation product in the solution should be a protonated form before workup. In the second step, the solution of the monoalkylation product was used for the next reaction without workup. Thus, the monoalkylation product was reacted with the N-acyliminium ion using the same flow microreactor system. When the residence time was set to 0.03 s, the reaction gave the dialkylation product with a yield of as small as 6 %. When the residence time was increased to 0.3 s, the reaction gave the dialkylation product with a yield of 56 %. This reveals that the second-stage reaction is slower than the first-stage reaction. It should be kept in mind that the second stage is also very fast because more than a half of the starting material reacted within 0.3 s at −78 °C, even if it is slower than the first stage.

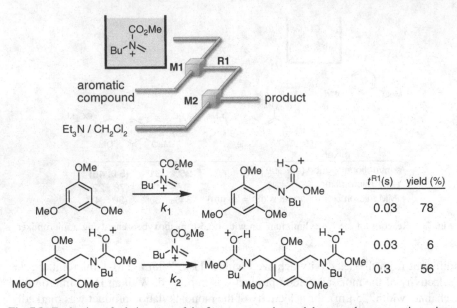

Fig. 7.8 Evaluation of relative rates of the first-stage reaction and the second-stage reaction using a flow microreactor system

7.6 CFD Simulations

Assuming that the first-stage reaction is faster than the second-stage reaction, simulations using computational fluid dynamics (CFD) were carried out (Fig. 7.9). The flow reactor contains laminar flows of **A** and **B**. Mixing is assumed to be

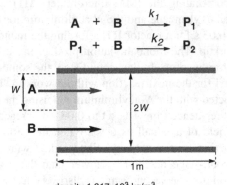

Fig. 7.9 CFD simulation of product selectivity of a competitive consecutive reaction

density 1.317×10^3 kg/m^3
diffusion coefficient 10^{-9} m^2/s
viscosity: 0.00119 Pa s;
initial concentrations of **A** and **B**: 0.01 mol/L.
linear flow speed 0.1 m/s

caused only by molecular diffusion. With no actual experimental values being available for various properties of the reaction components of the reaction of the N-acyliminium ion and the aromatic compounds required for the CFD simulations, the corresponding standard values shown in Fig. 7.9 were used. The reaction rate constants k_1 and k_2 were set so that k_1 is always 100 times the value of k_2. In other words, the first-stage reaction is 100 times faster than the second-stage reaction.

Simulations were conducted for four different cases: ideal mixing, and laminar flow widths of 100, 25, and 2.5 mm. The ideal mixing refers to mixing that produces a homogeneous solution immediately after the species enter the reaction tube.

Table 7.1 shows the results for the rate constant k_1 of 10^4. Under ideal mixing, the ratio of **P1** and **P2** is 97:3. This is the selectivity based on the kinetics. The selectivity under the laminar flow width of 100 μm is 75:25. The selectivity is improved to 95:5 for the laminar flow width of 25 μm. With the laminar flow width of 2.5 μm, the same selectivity as achieved with ideal mixing is achieved.

Table 7.2 shows the results for the rate constant k_1 of 10^5 Lmol^{-1}s^{-1} and the rate constant k_2 of 10^3 Lmol^{-1}s^{-1}. The ratio of the reaction rate constants is 1:100 as above. Under ideal mixing in which diffusion occurs instantaneously, increasing the reaction rate as a whole does not change the selectivity. However, the selectivity can change when the diffusion requires a finite time. With the laminar flow width of 100 μm, the products **P1** and **P2** are obtained at the ratio of substantially 1:1. With the laminar flow width of 25 μm, the selectivity improves to about 8:2, which still deviates greatly from the selectivity based on the kinetics. With the laminar flow width of 2.5 μm, the selectivity is 95:5, which is close to the selectivity based on the kinetics.

Table 7.1 Selectivity obtained by CFD simulation ($k_1 = 10^4$ L mol^{-1}s^{-1}, $k_2 = 10^2$ L mol^{-1}s^{-1})

Mixer	Laminar flow width (μm)	**P1** yield (%)	**P2** yield (%)	Selectivity **P1:P2**
Ideal mixing	–	94.6	2.7	97:3
(a)	100	60.6	19.7	75:25
(b)	25	89.6	4.5	95:5
(c)	2.5	94.7	2.6	97:3

Table 7.2 Selectivity obtained by CFD simulation ($k_1 = 10^5$ L mol^{-1}s^{-1}, $k_2 = 10^3$ L mol^{-1}s^{-1})

Mixer	Laminar flow width (μm)	**P1** yield (%)	**P2** yield (%)	Selectivity **P1:P2**
Ideal mixing	–	94.6	2.7	97:3
(a)	100	31.1	34.4	47:53
(b)	25	67.8	15.9	81:19
(c)	2.5	94.5	2.7	97:3

Table 7.3 Selectivity obtained by CFD simulation ($k_1 = 10^6$ L mol^{-1}s^{-1}, $k_2 = 10^4$ L mol^{-1}s^{-1}

Mixer	Laminar flow width (μm)	P1 yield (%)	P2 yield (%)	Selectivity P1:P2
Ideal mixing	–	94.6	2.7	97:3
(a)	100	14.5	42.7	25:75
(b)	25	36.1	31.9	53:47
(c)	2.5	90.5	4.7	95:5

Increasing the reaction rate further increases the effect of the laminar flow width, or the diffusion time. Table 7.3 shows the results of CFD simulations carried out with the rate constant k_1 of 10^6 Lmol^{-1}s^{-1} and the rate constant k_2 of 10^4 Lmol^{-1}s^{-1}. The selectivity under ideal mixing is the same as described above, whereas the selectivity is reversed to 25:75 with the laminar flow width of 100 μm, apparently due to disguised chemical selectivity. With the laminar flow width of 25 μm, the selectivity becomes substantially 1:1. With the laminar flow width of 2.5 μm, however, the selectivity is 95:5, which is close to the selectivity based on the kinetics.

The CFD simulations indicate that as the reaction rate increases, the laminar flow width or the diffusion time (mixing time) has more impact on the product selectivity and micromixing is more effective.

7.7 Controlling Other Reactions Using Micromixing

The same phenomenon is observed in various fast reactions with different reaction mechanisms, and in some cases, micromixing is reported to be very effective.

Like Friedel–Crafts reactions, halogenation such as bromination and iodination of aromatic compounds are classified as electrophilic aromatic substitution reactions. Bromine and iodine substituents are weakly electron-withdrawing groups, and introduction of such substituents causes a decrease in reactivity, and therefore, the first halogenation is faster than the second halogenation. However, the reactions in batch macroreactors suffer from the problem of disguised chemical selectivity, i.e, the formation of dihalogenated products.

The selectivity of bromination of 1,3,5-trimethoxybenzene in a flow microreactor [8] depends on the total flow rate, when a less effective micromixer such as Swagelok T-mixer is used (Fig. 7.10). Monobromination product and dibromination product are formed in almost 1:1 ratio at low flow rates. When a more effective micromixer such as IMVT cyclone mixer is used, the selectivity does not depend on the total flow rate and the highest monobromination/dibromination ratio of 100 is obtained.

The reactions of electron-rich aromatic compounds with I$^+$ generated by electrochemical oxidation of I$_2$ in CH$_3$CN also suffer from the problem of disguised chemical selectivity. The flow microreactor system incorporating a micromixer

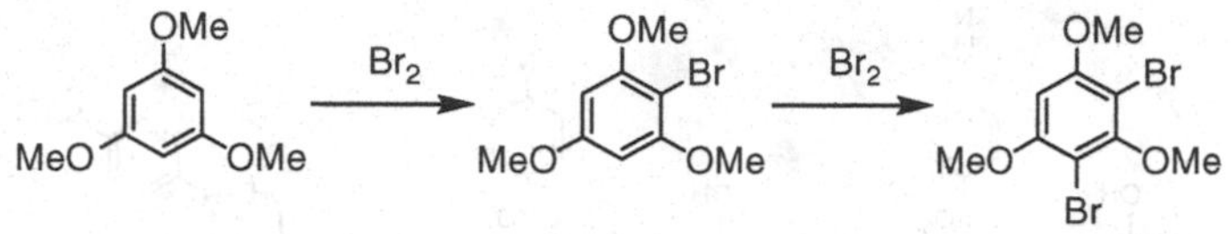

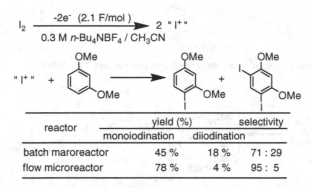

Fig. 7.10 Bromination of 1,3,5-trimethoxybenzene

$$I_2 \xrightarrow[\substack{0.3 \text{ M } n\text{-Bu}_4\text{NBF}_4 / \text{CH}_3\text{CN}}]{-2e^- \text{ (2.1 F/mol)}} 2 \text{ "I}^+\text{"}$$

reactor	yield (%)		selectivity
	monoiodination	diiodination	
batch maroreactor	45 %	18 %	71 : 29
flow microreactor	78 %	4 %	95 : 5

Fig. 7.11 Iodination of 1,3-dimethoxybenzene

enables highly selective formation of a product containing only one iodine atom (Fig. 7.11) [9, 10].

Micromixing improves the selectivity of the reaction between trimethoxyborane and a phenylmagnesium bromide (phenyl Grignard reagent) (Fig. 7.12) [11]. The nucleophilic attack of the Grignard reagent to boron replaces the methoxy groups on the boron by the phenyl group. Trimethoxyborane contains three methoxy groups, and thus, up to three phenyl groups can be attached to the boron atom. A boron compound containing one phenyl group produced in the first-stage reaction is less reactive than trimethoxyborane, which is the starting material. Thus, the

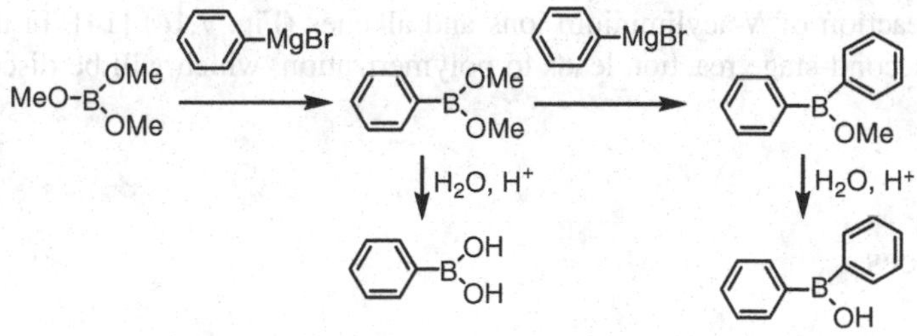

batch macroreactor	70.6%	13.8%
flow microreactor	93.9%	0.6%

Fig. 7.12 Product selectivity of the reaction of phenylmagnesium bromide with trimethoxyborane

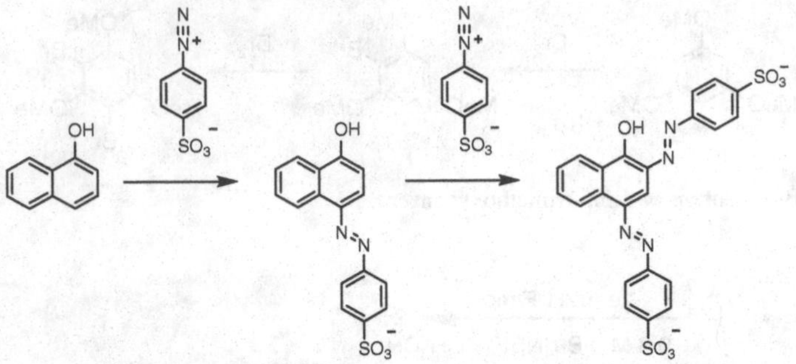

Fig. 7.13 Diazocoupling of 4-sulfobenzenediazonium ion and 1-naphthol

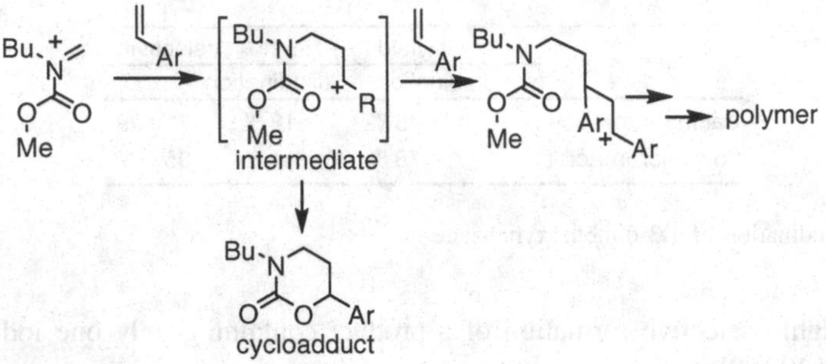

Fig. 7.14 [4 + 2] Cycloaddition of an *N*-acyliminium ion and styrene derivatives

reaction can be kinetically stopped at the first stage. However, such reaction conducted in a batch macroreactor produces a substantially large amount of product containing two phenyl groups as a by-product. The flow microreactor system incorporating a micromixer enables extremely highly selective formation of a product containing only one phenyl group.

The effect of micromixing was also observed for diazo coupling of electron-rich aromatic compounds and diazonium ions (Fig. 7.13) [12, 13] and [4 + 2] cycloaddition reaction of *N*-acyliminium ions and alkenes (Fig. 7.14) [14]. In the latter case, the second-stage reaction leads to polymerization, which will be discussed in Chap. 10.

References

1. P. Rys, Acc. Chem. Res. **9**, 345 (1976)
2. P. Rys, Angew. Chem. Int. Ed. Engl. **16**, 807 (1977)
3. S. Suga, A. Nagaki, J. Yoshida, Chem. Commun. 354 (2003)

4. A. Nagaki, M. Togai, S. Suga, N. Aoki, K. Mae, J. Yoshida, J. Am. Chem. Soc. **127**, 11666 (2005)
5. J. Yoshida, S. Suga, Chem. Eur. J. **8**, 2651 (2002)
6. J. Yoshida, K. Kataoka, R. Horcajada, A. Nagaki, Chem. Rev. **108**, 2265 (2008)
7. J. Yoshida, Y. Ashikari, K Matsumoto, T. Nokami, J. Synth. Org. Chem. Jpn. **71**, 1136 (2013)
8. K.J. Hecht, A. Kolbl, M. Kraut, K. Schubert, Chem. Eng. Technol. **31**, 1176 (2008)
9. K. Midorikawa, S. Suga, J. Yoshida, J. Chem. Commun. 3794 (2006)
10. K. Kataoka, Y. Hagiwara, K. Midorikawa, S. Suga, J. Yoshida, Org. Process Res. Dev. **12**, 1130 (2008)
11. V. Hessel, C. Hofmann, H. Löwe, A. Meudt, S. Scherer, F. Schönfeld, B. Werner, Org. Process Res. Dev. **8**, 511 (2004)
12. K. K. Cotı, Y. Wang, W.-Y. Lin, C.-C. Chen, Z. T. F. Yu, K. Liu, C. K.-F. Shen, M. Selke, A. Yeh, W. Lu, H.-R. Tseng, Chem. Comm. 3426 (2008)
13. H. Hisamoto, T. Saito, M. Tokeshi, A. Hibara, T. Kitamori, Chem. Commun. 2662 (2001)
14. S. Suga, A. Nagaki, Y. Tsutsui, J. Yoshida, Org. Lett. **5**, 945 (2003)

Chapter 8
Flash Chemistry

Abstract Flash chemistry is a field of chemical synthesis using flow microreactors where extremely fast reactions are conducted in a highly controlled manner to produce desired compounds with high selectivity. The concept of flash chemistry can be applied to various types of reactions. Chemical reactions that are practically impossible in batch macroreactors should be made possible by flash chemistry. Flash chemistry can make strong impact on industry.

8.1 Introduction

In Chaps. 3–6, we have learned that the residence time control allows highly unstable intermediates to form in short times and react with other compounds before decomposing or isomerizing (high-resolution reaction time control). In Chap. 7, we have learned that the speed of mixing directly affects the product selectivity of some fast reactions such as competitive consecutive reactions and that extremely fast mixing using a micromixer is indispensable for the control of such reactions to improve the product selectivity. We may now realize that these features of flow microreactors would enable chemical synthesis that cannot be done in batch. Such chemical synthesis using a flow microreactor is called "flash chemistry" [1–5]. Flash chemistry allows chemical transformations that are very difficult or practically impossible to conduct using conventional batch macroreactors. This chapter describes the concept and some applications of flash chemistry, although some examples have already shown in the previous chapters.

8.2 What Is Flash Chemistry?

In the first dedicated book [6] to this exciting field, flash chemistry is defined as follows. Flash chemistry is a field of chemical synthesis using flow microreactors where extremely fast reactions are conducted in a highly controlled manner to

© The Author(s) 2015
J. Yoshida, *Basics of Flow Microreactor Synthesis*,
SpringerBriefs in Molecular Science, DOI 10.1007/978-4-431-55513-1_8

produce desired compounds with high selectivity. Although the definition might change with progress in the technologies related to this field, this definition still holds good. In flash chemistry, the reaction time usually ranges from milliseconds to seconds, much shorter than that in flask chemistry. It is practically impossible to conduct such fast reactions in a conventional batch macroreactor such as a flask on a preparative scale. Flash chemistry would provide a new method for chemical science, where extremely fast reactions that have never been performed in a flask can be developed. Flash chemistry may lead to the creation of new materials and new biologically active compounds. We should keep in mind that the typical reaction times in flash chemistry might change with progress in related technologies including microfluidic technology. Therefore, there should be many scientific challenges that can be accomplished using flash chemistry in the future.

The word "flash" has been used in the history of chemistry for many years. For example, flash vacuum pyrolysis [7, 8] is a well-known technique that has been used for chemical synthesis at high temperatures. Flash laser photolysis [9, 10] serves as a powerful method for generating reactive species in a very short time and has been used for mechanistic studies of extremely fast light induced chemical processes which are complete within milliseconds or less. Flash chromatography [11] is one of the most popular techniques for separating and purifying compounds in organic chemistry laboratories. It should be noted, therefore, that flash chemistry is a new field of chemical synthesis but that the word "flash" is very common in chemistry.

8.3 Some Examples of Flash Chemistry

There are many possibilities in flash chemistry. Although some examples of flash chemistry have already shown in the previous chapters, the following examples also demonstrate the power of flash chemistry.

Swern–Moffatt-type oxidation of alcohols to carbonyl compounds is widely used in laboratory synthesis, but it requires very low temperatures such as −50 °C or below, because the intermediates are very unstable. Such a requirement causes severe limitations in the industrial use of this highly useful reaction. The use of a flow microreactor system consisting of three micromixers (M1, M2, and M3) and three microtube reactors (R1, R2, and R3) solves the problem. By shortening the residence time in R1 to 10 ms using a flow microreactor system enables generation of such intermediates and their reaction in the next step before decomposition (Fig. 8.1). Thus, the reaction can be performed at room temperature (R1 and R2) avoiding cryogenic conditions [12]. By premixing dimethylsulfoxide and an alcohol, the reaction can be performed at higher temperatures such as 70 °C [13].

An integrated flow microreactor system fabricated on a polymer–glass chip consisting of two T-junctions (T1 and T2) and two microchannel reactor (R1 and R2) can also be effectively used for flash chemistry (Fig. 8.2) [14]. Halogen–Lithium exchange reaction of o-bromophenyl isocyanide with n-BuLi generates a

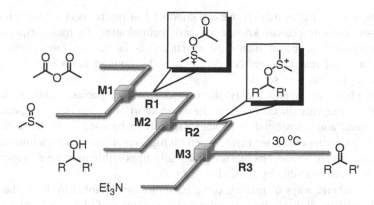

Fig. 8.1 Swern–Moffatt oxidation of alcohols using a flow microreactor system

Fig. 8.2 Synthesis of tryptanthrin using the pressure-tolerant PVSZ-glass microreactor

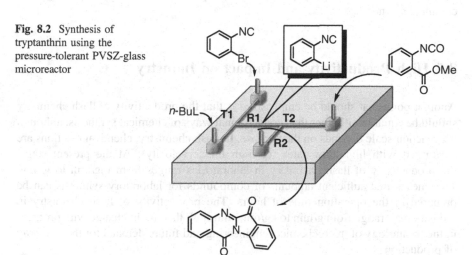

highly reactive aryllithium species bearing a isocyanide group, which is immediately reacted with phenyl isocyanate bearing an alkoxycarbonyl group to produce tryptanthrin, a drug compound at room temperature (Fig. 8.2).

8.4 Scientific Merits

Some people may think that flask chemistry is powerful enough and reactions that are complete in hours fit well for human beings. Why is flash chemistry needed? The most appropriate answer to this question is because we can just do it. Reaction times at the molecular level are generally several hundred femtoseconds or picoseconds. Although the reaction times in flash chemistry are much longer than the scientific limit of reaction times, we can conduct fast reactions at a natural reaction rate based on our present knowledge and technologies. In many aspects, fast reactions would be better than slow reactions as far as we can control them. Conducting fast reactions without decelerating by cooling or using additives are time-, energy-, and cost-saving.

Flash chemistry makes highly short-lived reactive species usable as starting materials or reagents for chemical synthesis. In short, unstable reactive species can be generated and transferred to another location to be used in the next reaction before they decompose or isomerize based on high-resolution reaction time control. Therefore, chemical reactions that are practically impossible in batch macroreactors should be made possible by flash chemistry.

New synthetic ways of making complex molecules enabled by flash chemistry would lead to creation of new molecules and compounds of biological activities and functions as materials. Thus, flash chemistry introduces a new paradigm in chemical science.

8.5 High Productivity and Impact on Industry

Another point that should be emphasized is that the productivity of flash chemistry should be equal to or higher than conventional ways of chemical synthesis, although the reaction scale depends on the purpose. In flash chemistry, chemical reactions are performed with high flow rates (or high linear velocity). At the present stage, the productivity of flash chemistry in laboratories ranges from mg/min to g/min. This means that sufficient amounts of compounds for laboratory synthesis can be prepared in the operation time of hours. The productivity of flash chemistry in industry may range from g/min to kg/min, although this might change with progress in the technology of microchemical engineering and future demand for the new way of production.

Also, the residence time in flash chemistry ranges from milliseconds to seconds. This means that the operation of the reaction can be started very quickly to get a steady state and that the operation can be stopped in very short time. This feature is also beneficial from a viewpoint of industrial production.

Thus, flash chemistry would make strong impact on the pharmaceutical, agro-chemical, fine-chemical, petro-chemical, polymer, inorganic, and fragrance industries.

References

1. A. Nagaki, M. Togai, S. Suga, N. Aoki, K. Mae, J. Yoshida, J. Am. Chem. Soc. **127**, 11666 (2005)
2. J. Yoshida, Chem. Commun. 4509 (2005)

3. J. Yoshida, A. Nagaki, T. Yamada, Chem. Eur. J. **14**, 7450 (2008)
4. J. Yoshida, Chem. Rec. **10**, 332 (2010)
5. J. Yoshida, Y. Takahashi, A. Nagaki, Chem. Commun. **49**, 9896 (2013)
6. J. Yoshida, *Flash chemistry: Fast organic synthesis in microsystems* (Wiley-Blackwell, Hoboken, 2008)
7. V. Boekelheide, Acc. Chem. Res. **13**, 65 (1980)
8. P.W. Rabideau, A. Sygula, Acc. Chem. Res. **29**, 235 (1996)
9. P.K. Das, Chem. Rev. **93**, 119 (1993)
10. A.J. Kresge, J. Phys. Org. Chem. **11**, 292 (1998)
11. W.C. Still, M. Kahn, A. Mitra, J. Org. Chem. **43**, 2923 (1978)
12. T. Kawaguchi, H. Miyata, K. Ataka, M. Kazuhiro, J. Yoshida, Angew. Chem. Int. Ed. **44**, 2413 (2005)
13. P.J. Nieuwland, K. Koch, N. van Harskamp, R. Wehrens, J.C.M. van Hest, F.P.J.T. Rutjes, Chem. Asian J. **5**, 799 (2010)
14. W. Ren, H. Kim, H.-J. Lee, J. Wang, H. Wang, D.-P. Kim, Lab Chip **14**, 4263 (2014)

Chapter 9
Space Integration of Reactions

Abstract A system integrating multiple flow microreactors that are connected to one another allows multiple reactions to be carried out continuously, or allows space integration of reactions. Flow microreactors that can set short residence times enable integration of reactions involving unstable short-lived reactive species as intermediates. Space integration of not only reactions of the same type but also of reactions of different types can be performed.

9.1 Introduction

We have learned in Chaps. 4–6 that high-resolution reaction time control using a flow microreactor system allows unstable intermediates to be generated in a short time and be usable in reactions with other compounds without decomposing. An integrated system consisting of micromixers and flow microreactors that are connected to one another will allow multiple reactions to be carried out continuously. This method is called space integration of reactions. The method also allows integration of reactions involving unstable intermediates by setting shorter residence times. This chapter describes the principle and examples of space integration of reactions.

9.2 Integrating Reactions

Synthetic chemistry has achieved many advances by formulating reactions in which one molecule is precisely transformed into another molecule to allow efficient and selective synthesis of the molecules of a desired product. The reaction conditions have been optimized individually for different reactions. However, producing a desired chemical compound with a one-step reaction is often difficult, and it usually requires a combination of multiple reactions. Some recent medical product may

© The Author(s) 2015
J. Yoshida, *Basics of Flow Microreactor Synthesis*,
SpringerBriefs in Molecular Science, DOI 10.1007/978-4-431-55513-1_9

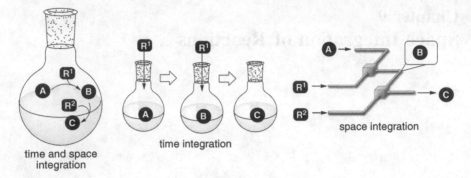

Fig. 9.1 Classification of the way of integrating reactions

require 60 or more steps for its synthesis. In contrast to this, organisms use a wide variety of chemical reactions occurring in a cooperative, coherent and well-ordered manner, and efficiently produce required compounds, whereas synthetic chemistry still remains at an elementary level. Synthetic chemistry using reaction integration combines multiple chemical reactions over time and space to formulate efficient methods for molecular transformation, aiming to precisely and promptly synthesize organic molecules having desired biological activity or physical functionality. As shown in Fig. 9.1, such methods for integration can be classified into three types: time and space integration of reactions, time integration of reactions, and space integration of reactions [1, 2].

Time and space integration of reactions are the method of integrating reactions occurring at one time in one reactor in a cooperative and interrelated manner, or in a concerted manner. Examples include domino reactions or tandem reactions. Another important example is a concerted catalytic reaction, which involves multiple catalysts or a single catalyst with multiple functions. This approach allows reactive species to be trapped promptly by the coexisting molecules and thus allows the integration of reactions involving extremely short-lived reactive intermediates. However, this approach requires a fixed sequence of reactions, which is primarily determined by the combination of chemical compounds. The sequence cannot be changed. Time and space integration of reactions thus has lower flexibility in its design and execution of reactions than the other two approaches.

Time integration of reactions is the method of integrating a series of reactions occurring successively in one reactor. The reactions occur sequentially on the time axis in the same reactor. This method has been used conventionally as a one-pot sequential synthesis. After one reaction is complete, a substrate(s) or a reagent(s) or a catalyst(s) is added to cause another reaction. This approach cannot be used for reactions in which unstable species, such as short-lived reactive species, occur as intermediates. Time integration of reactions permits easy changes in the sequence of compounds to be added and thus is more flexible in its design and execution of reactions than the time and space integration approach.

Space integration of reactions is the method of integrating reactions occurring in different reactors that are arranged spatially. To allow faster transportation of compounds between the different reactors, a continuous flow system is more advantageous than a batch system. In the book titled "Integrated Chemical Systems" (Wiley, 1994), A.J. Bard proposed and predicted an integrated chemical synthesizer, which has now come to the existence with this approach. Like the time integration approach, space integration of reactions also permits easy changes in the sequence of compounds to be added and is thus highly flexible. Using flow microreactors and setting short residence times to allow use of species having a lifetime of a millisecond, this approach can integrate reactions involving short-lived reactive intermediates. Although the space integration approach is being studied actively only a few research examples are reported for the approach applied to reactions involving short-lived reactive intermediates.

This chapter describes specific examples of the space integration approach using flow microreactors, particularly the approach based on flash chemistry involving short-lived reactive intermediates.

9.3 Synthesizing *o*-Disubstituted Benzenes by Space Integration of Two Organolithium Reactions [3, 4]

o-Bromophenyllithium generated by the Br–Li exchange reaction of *o*-dibromobenzene with *n*-BuLi is known to rapidly undergo elimination of LiBr to give benzyne. If *o*-bromophenyllithium can readily react with various electrophiles before it undergoes elimination, this process should enable synthesis of various substituted benzenes containing a bromo group at the *ortho* position. To reduce the elimination and enable the reaction with an electrophile in a batch reactor, the compound reportedly needs to be placed at an extremely low temperature of −110 °C [5]. If this reaction is carried out in a batch reactor at −78 °C, no intended product is obtained. This is presumably because the batch reactor cannot allow reactions to complete in a short time unlike the flow microreactor system and thus fails to prevent *o*-bromophenyllithium from decomposing into benzyne.

This transformation, however, was successfully carried out using a flow microreactor system. *o*-Bromophenyllithium was allowed to react with various electrophiles under optimum conditions, or at the temperature of −78 °C and the residence time of 0.82 s.

The product is a bromobenzene derivative with a benzene ring to which one electrophile has been added. The remaining Br group can further undergo a Br–Li exchange reaction if *n*-BuLi is added, and the resultant aryllithium species can react with another electrophile.

An integrated flow microreactor system consisting of four micromixers and four flow microreactors shown in Fig. 9.2 can perform space integration of such reactions. In micromixer **M1**, *o*-dibromobenzene is mixed with *n*-BuLi. In reactor **R1**,

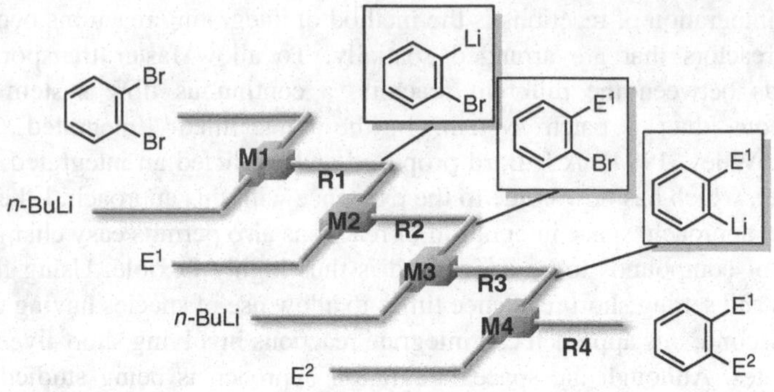

Fig. 9.2 Synthesis of *o*-disubstituted benzenes from *o*-dibromobenzene by space integration of reactions

the Br–Li exchange reaction is carried out. The resultant *o*-bromophenyllithium is then mixed with a first electrophile (E1) in micromixer **M2**. The generated intermediate then reacts with *n*-BuLi to undergo a Br–Li exchange reaction again in **M3** and **R3**. The generated aryllithium species then reacts with a second electrophile (E2) in **M4** and **R4**, affording the corresponding *o*-disubstituted benzene.

Figure 9.3 shows the results obtained with various electrophiles. Importantly, each step requires the setting of the corresponding optimum residence time and temperature, because the initially generated aryllithium species and the subsequently generated aryllithium species would differ in stability and reactivity. As described above, the initially generated *o*-bromophenyllithium is so unstable that it

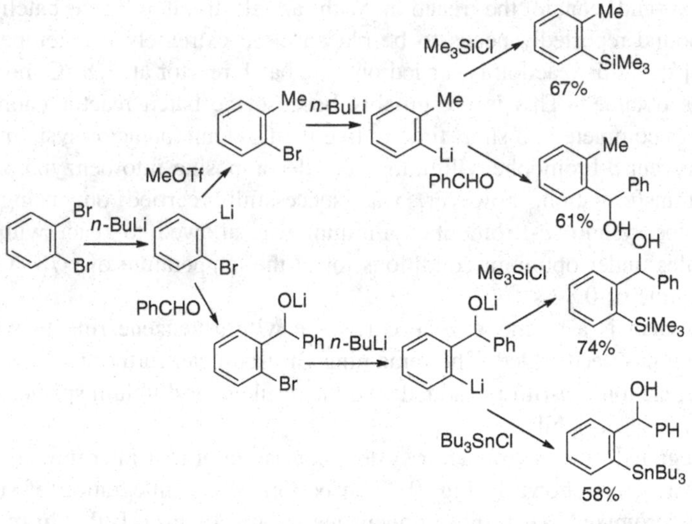

Fig. 9.3 Synthesis of various *o*-disubstituted benzenes from *o*-dibromobenzene

can readily decompose into benzyne. Therefore, *o*-bromophenyllithium intermediate needs to be reacted with various electrophiles at −78 °C with a residence time
of 0. 82 s. The subsequent Br–Li exchange reaction can be carried out at 0 °C,
because the aryllithium species to be generated is relatively stable because it does
not contain bromine at the *ortho* position, which would undergo elimination.
Notably, trimethylsilyl triflate should be used as E1 for the reaction of *o*-bro-
mophenyllithium, whereas chlorotrimethylsilane with lower reactivity can be used
as an E2. It is also noteworthy that chlorotributylstannane cannot be used as an E1,
because the resulting arylstannane would undergo a Sn–Li exchange reaction in the
second lithiation process.

9.4 Synthesizing TAC-101 by Integration of Three Organolithium Reactions [6]

TAC-101 (4-[3,5-bis(trimethylsilyl)benzamido] benzoic acid) has attracted attention as a compound with antitumor activity. An ester of this compound can be
readily synthesized by using 1,3,5-tribromobenzene as a starting material, and
repeating three times the Br–Li exchange followed by the reaction of the generated
aryllithium species with an electrophile (Fig. 9.4). For these reactions, a system
integrating six micromixers and six flow microreactors is used. When each step uses
an optimum residence time, the steps together require a total residence time of 13 s.

The all reactions can be carried out at 0 °C. A laboratory flow microreactor
system can produce 100–200 mg of the product per minute. The flow microreactor
system easily allows the synthesis of an unsymmetrically substituted compound
containing silyl groups with different substituents as well.

9.5 Space Integration of Halogen–Lithium Exchange Reaction and Cross-Coupling Reaction [7]

Although integration of multiple reactions of the same type has been described
above, reactions of completely different types can also be integrated. Preparing an
aryllithium species through a halogen–lithium exchange reaction and performing
space integration of this reaction and the Pd-catalyzed cross-coupling (Murahashi
coupling) [5] reaction using the prepared species will now be described.

The halogen–lithium exchange reaction of an aryl halide using a flow microreactor system has been discussed in the previous chapters. If the aryllithium
species can be used in a coupling reaction catalyzed by a Pd complex in the flow
microreactor system, this would provide an efficient and effective method of synthesis. However, the two reactions cannot simply be carried out sequentially. For

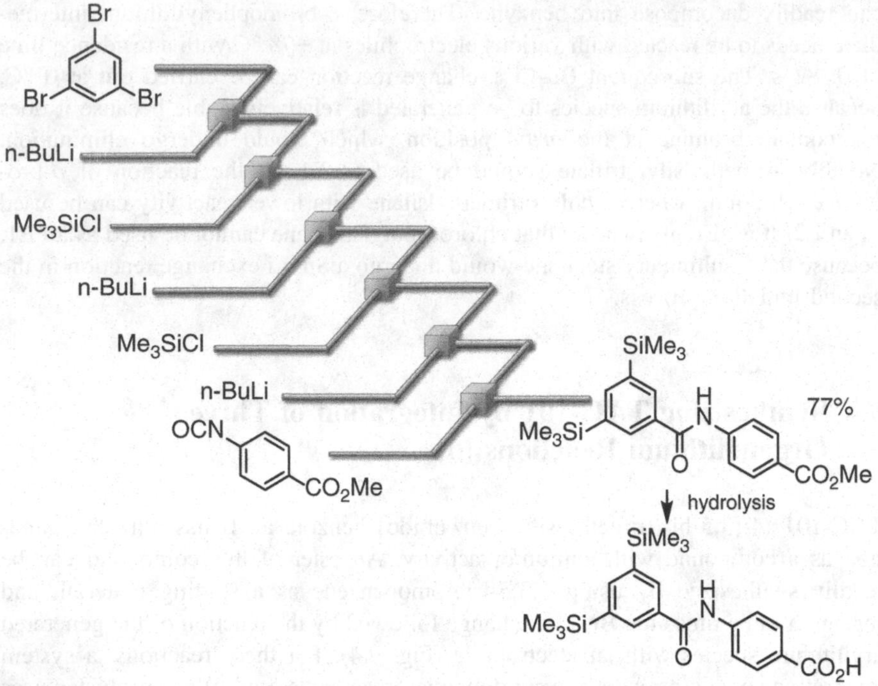

Fig. 9.4 Synthesis of TAC-101 by space integration of organolithium reactions

example, Br–Li exchange of an ArBr using *n*-BuLi would produce an aryllithium species (ArLi), whereas this reaction also produces *n*-butyl bromide (*n*-BuBr) in the same quantity as the aryllithium species. Typically, a Pd-catalyzed cross-coupling reaction of ArLi with an aryl halide (Ar'X), which is subsequently added as a coupling partner, is slow, and the reaction of ArLi with *n*-BuBr would thus be predominant. There is another possibility. When the cross-coupling is slow, ArLi can undergo a halogen–lithium exchange reaction with Ar'X to give Ar'Li. This leads to a product derived from the reaction of Ar'Li and *n*-BuBr as well as the homocoupling products, i.e, Ar–Ar and Ar'–Ar'. A specific example will be described with reference to Fig. 9.5. To solve such problems with the integration of reactions, the Pd-catalyzed cross-coupling reaction may need acceleration to suppress the side reactions. The use of PEPPSI-SIPr containing a carbene ligand as a catalyst was found to accelerate the cross-coupling reaction, which would proceed faster than the side reactions. This enables the synthesis of the desired cross-coupling product with a relatively high yield.

The Br–Li exchange reaction of *p*-methoxybromobenzene and the cross-coupling reaction of the resulting *p*-methoxyphenyllithium with bromobenzene, catalyzed by PEPPSI-SIPr, were carried out using a flow microreactor system shown in Fig. 9.6,

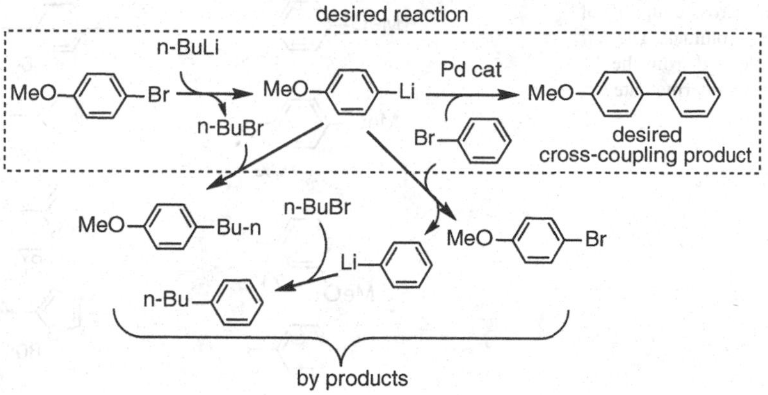

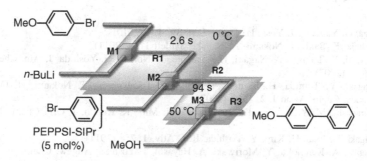

Fig. 9.5 Integration of Br–Li exchange of *p*-methoxybromobenzene and Pd-catalyzed cross-coupling with bromobenzene

Fig. 9.6 Space integration of Br–Li exchange of *p*-methoxybromobenzene and Murahashi coupling with bromobenzene

enabling the synthesis of a desired cross-coupling product with a yield of 93 %. Under these conditions, however, the cross-coupling requires a reaction time (94 s at 50 °C in **R2**), which is much longer than the reaction time taken by the Br–Li exchange reaction (2.6 s at 0 °C in **R1**). Thus, an unstable aryllithium species containing an ester or a ketone carbonyl group cannot be used in the cross-coupling reaction. A catalyst with much higher activity needs to be developed.

Although the present method suffers from the functional group compatibility, various compounds can be synthesized from two different aryl bromides as shown in Fig. 9.7. Heteroaryllithiums, which are readily prepared by H–Li exchange reaction, can be used for the coupling to give heteroaryl–aryl and heteroaryl–heteroaryl coupling products.

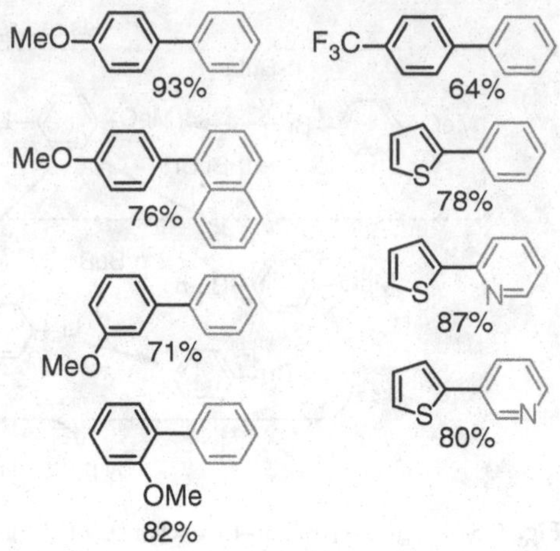

Fig. 9.7 Cross-coupling of
two aryl bromides. The *left
part* is derived from the
aryllithium intermediate

MeO—⟨ ⟩—⟨ ⟩   F₃C—⟨ ⟩—⟨ ⟩
        93%   64%

MeO—⟨ ⟩—⟨naphthalene⟩
        76%

⟨thiophene⟩—⟨ ⟩
                    78%

⟨ ⟩—⟨ ⟩
MeO   71%

⟨thiophene⟩—⟨pyridine⟩
                    87%

⟨ ⟩—⟨ ⟩
OMe
82%

⟨thiophene⟩—⟨pyridine⟩
                    80%

References

1. S. Suga, D. Yamada, J. Yoshida, Chem. Lett. **39**, 404 (2010)
2. J. Yoshida, K. Saito, T. Nokami, A. Nagaki, Synlett 1189 (2011)
3. H. Usutani, Y. Tomida, A. Nagaki, H. Okamoto, T. Nokami, J. Yoshida, J. Am. Chem. Soc. **129**, 3046 (2007)
4. A. Nagaki, Y. Tomida, H. Usutani, H. Kim, N. Takabayashi, T. Nokami, H. Okamoto, J. Yoshida, Chem. Asian J. **2**, 1513 (2007)
5. S. Murahashi, M. Yamamura, K. Yanagisawa, N. Mita, K. Kondo, J. Org. Chem. **44**, 2409 (1979)
6. A. Nagaki, K. Imai, H. Kim, Y. Yoshida, RSC Adv. **1**, 758 (2011)
7. A. Nagaki, A. Kenmoku, Y. Moriwaki, A. Hayashi, J. Yoshida, Angew. Chem. Int. Ed. **49**, 7543 (2010)

Chapter 10
Polymerization Using Flow Microreactor System

Abstract Fast mixing and fast heat transfer of a flow microreactor enable living cationic polymerization of vinyl ethers without any additives to stabilize an active propagating polymer end. Such polymerization involves no equilibrium between the active species and the dormant species, and thus proceeds fast and is complete in a short time. Setting a short residence time enables the propagating polymer end to be usable in a subsequent reaction such as polymerization of a different monomer to make a block copolymer and end functionalization. The flow microreactor enables anionic polymerization of styrene or methacrylate monomers at higher temperatures than in a conventional batch reactor. The propagating polymer end in anionic polymerization can be used in a subsequent reaction, allowing synthesis of a polymer containing a functional group at its end or allowing synthesis of a block copolymer.

10.1 Introduction

In Chap. 8, we have learned that a flow microreactor system incorporating a micromixer is useful in controlling competitive consecutive reactions. The ultimate reaction system in which reactions occur in chains, or consecutively, is polymerization. This chapter describes how we can exploit the advantages of the flow microreactor system in controlling the molecular weight or molecular weight distribution in polymerization reactions, including cationic polymerization and anionic polymerization.

10.2 Controlling Competitive Consecutive Reactions and Chain-Growth Polymerization

Polymerization reactions can be basically categorized into two types: chain-growth (or addition) polymerization and step-growth (or condensation) polymerization. In chain-growth polymerization, an active species resulting from the initiation process

© The Author(s) 2015
J. Yoshida, *Basics of Flow Microreactor Synthesis*,
SpringerBriefs in Molecular Science, DOI 10.1007/978-4-431-55513-1_10

reacts with a monomer to generate a new active species, which then reacts with another monomer to generate a active species with a longer chain length. This propagating polymer chain always contains an active species at its end, which then reacts with a monomer to form a longer polymer chain. Chain-growth polymerization can be classified as cationic polymerization, radial polymerization, or anionic polymerization depending on the type of the active species. For step-growth polymerization, each step of polymerization produces a stable polymer molecule, which reacts with a monomer molecule or with a different polymer molecule to produce a polymer molecule with a longer chain length.

This chapter focuses on chain-growth polymerization and explores how we can use the flow microreactor system in controlling polymerization reactions, by analogy with controlling short-lived active species as well as controlling competitive consecutive reactions (Fig. 10.1). In Chap. 7, we have learned about controlling competitive consecutive reactions by 1:1 fast mixing using a micromixer to generate a product reacted at 1:1 ratio with high selectivity based on the kinetics. If fast mixing at $1:n$ ratio of the initiator to the monomer in polymerization can simply generate a polymer reacted at $1:n$ ratio, we should be able to control the molecular weight. The molecular weight distributes based on statistics, and thus, the molecular weight distribution is also important.

To control the molecular weight, all polymer chains need to start propagating at the same time. To enable this, the initiation reaction needs to be faster than the propagating reaction. However, when the reaction is very fast and is faster than mixing, the problem of disguised chemical selectivity occurs as we learned in Chap. 7. This can be particularly problematic in batch macroreactors. The flow microreactor system incorporating a micromixer would be effective in solving this problem.

Although chain-growth polymerization usually generates much heat, such heat of polymerization can be efficiently removed by the large specific surface area of the microreactor. This is an advantage of polymerization in a flow microreactor. Additionally, the residence time control using the flow microreactor system enables effective use of short-lived, unstable active species such as carbocations or carboanions at the ends of propagating polymers.

Fig. 10.1 Competitive consecutive reaction and chain-growth polymerization

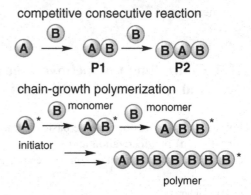

10.3 Chemistry of Cationic Polymerization

Before discussing controlling cationic polymerization using the flow microreactor system, let us briefly touch on chemistry of cationic polymerization. Cationic polymerization is one type of chain-growth polymerization that uses cationic species as active species. In the initiation step, a cation as an initiator is added to a vinyl monomer to generate a new cationic species as shown in Fig. 10.2. In the propagation step, the polymer chain propagates as the cation is added to another vinyl monomer, and the resultant cationic species is added to still another vinyl monomer. If the polymerization proceeds simply in this manner, the average molecular weight would be determined simply by the ratio of the initiator and the monomer.

However, controlling the molecular weight or controlling the molecular weight distribution is complicated and difficult because of chain transfer (Fig. 10.3). Every reaction of the cation at the propagating polymer end with a vinyl monomer might cause deprotonation at the β position, allowing the end of the polymer chain to have an alkene structure and stop propagating. A vinyl monomer receiving the proton forms the initiation end of a new polymer, at which the propagation reaction starts again. Such chain transfer makes it impossible to control the molecular weight simply based on the ratio of the initiator and the monomer. The chain transfer is an elimination reaction, which commonly occurs more easily at higher temperatures than other reactions. For the competition between S_N2 and E2 reactions, the E2 reaction is known to occur more easily at high temperatures. Thus, from this analogy, rapidly removing the heat of polymerization and preventing the temperature increase is important in suppressing the chain transfer reaction.

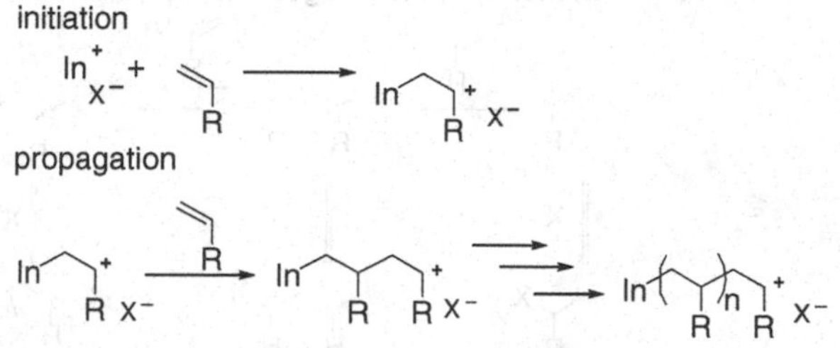

Fig. 10.2 Initiation and propagation of cationic polymerization

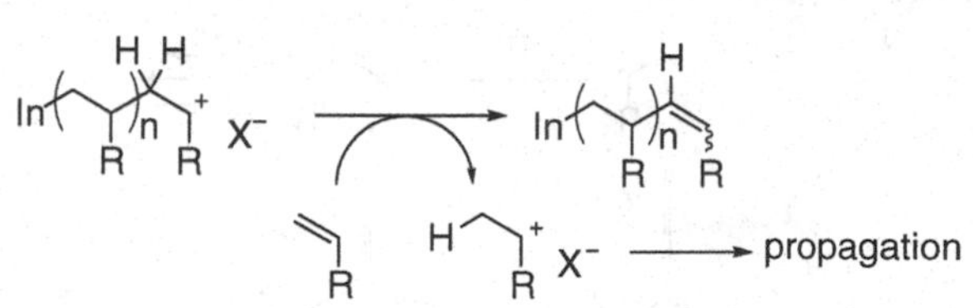

Fig. 10.3 Chain transfer reaction in cationic polymerization

Living cationic polymerization has been developed to control the molecular weight and the molecular weight distribution [1, 2, 3]. Living cationic polymerization is a polymerization reaction conducted in the presence of a weak nucleophile or a counter anion with weak nucleophilicity to form an equilibrium between the cationic propagating end (active species) and a relatively stable dormant species resulting from the reaction between the active species and the nucleophile (Fig. 10.4). This equilibrium is predominantly shifted to the dormant species, and thus, the active species constitutes a significantly lower concentration. Making this state slows the entire reaction down and suppresses the chain transfer. In this state, polymerization generates only a small amount of heat per unit time and thus is less likely to cause local temperature increase. This seems to be responsible for suppressing chain transfer reactions. The resultant polymer will have a narrow distribution of molecular weights.

In general, ideal living polymerization will occur under the conditions described below:

1. All polymer chains start propagating at the same time.
2. Each polymer chain reacts with a monomer at the same reaction rate.
3. The polymerization involves no reactions that would form inactive polymer chains, or thus involves no termination or no chain transfer reaction.
4. The polymerization stops only when the system includes no more monomers.
5. The polymerization can restart when the same or a different monomer is added to the system. Adding a different monomer will produce a block copolymer.
6. Adding a reagent that can stop polymerization would terminate the polymerization and form inactive polymer chains.

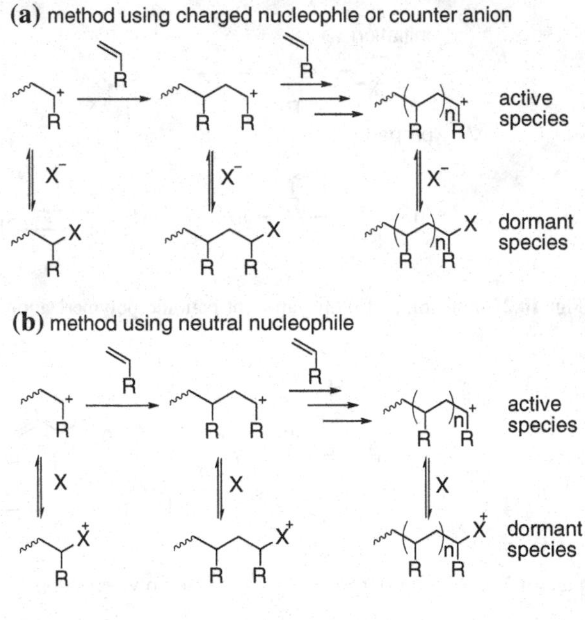

Fig. 10.4 Living cationic polymerization

(**a**) method using charged nucleophle or counter anion

(**b**) method using neutral nucleophile

These conditions are applicable not only to cationic polymerization but also to radical polymerization and anionic polymerization. The polymer produced under these conditions being satisfied will have a narrow distribution of molecular weights, with the molecular weights being controlled. These conditions can also allow the polymer end to contain a functional group, or can also allow block copolymerization.

10.4   Flash Cationic Polymerization

Polymerization of vinyl ethers initiated by an N-acyliminium ion pool has demonstrated that the molecular weight and the molecular weight distribution can be controlled by using the flow microreactor system (Fig. 10.5) [4, 5]. An N-acyliminium ion generated and accumulated by the cation pool method was used as an initiator, which was mixed with a vinyl ether at high speed using a micromixer. The polymerization proceeded in a flow microreactor and was complete within a short residence time. An amine was then introduced using a micromixer to terminate the polymerization. Thus, this polymerization can be called flash polymerization.

Such polymerization was conducted with the ratio of the monomer to the initiator being varied. The molecular weight increased in proportion to the ratio of the monomer to the initiator as shown in Fig. 10.6. The average molecular weight was close to a value estimated from the monomer-to-initiator ratio. This reveals that fast mixing at a ratio of 1:n successfully gives a polymer product reacted at 1:n ratio. The index Mw/Mn of the molecular weight distribution (the ratio of the weight average molecular weight (Mw) to the number average molecular weight (Mn)) was as small as 1.14.

In the flow microreactor, heat transfers at high speed and the reaction can proceed uniformly at a low temperature. This seemingly suppressed the chain transfer reaction. If the same polymerization is conducted in a normal batch macroreactor, the ratio Mw/Mn would be higher than 2. This may be because the slow mixing by stirring in the batch macroreactor prevents the propagation from starting

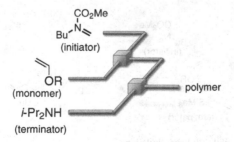

Fig. 10.5 Flash cationic polymerization of a vinyl ether using an N-acyliminium ion as an initiator in a flow microreactor

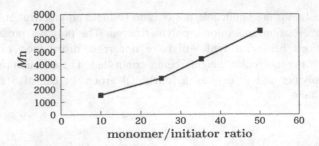

Fig. 10.6 Relationship between the molecular weight and the monomer/initiator ratio in cationic polymerization in a flow microreactor system

at the same time. Additionally, the batch macroreactor cannot remove heat of polymerization sufficiently rapidly, allowing local temperature increase and causing chain transfer reactions to occur easily.

For living polymerization, the end of the propagating polymer should remain active and living. To determine this phenomenon, the polymerization was terminated by using allyltrimethylsilane (Fig. 10.7). This should cause the polymer end to contain an allyl group. In the polymerization using the flow microreactor system, allyltrimethylsilane was introduced from a second micromixer, and then, the resultant polymer was determined to contain an allyl group at its end by the ^{1}H NMR spectroscopy of the polymer (Fig. 10.8). Additionally, the proton integral ratio of the methyl group of carbamate at the initiation end and alkene at the termination end also reveals that the initiation end and the termination end exist at the ratio of substantially 1:1. This finding also supports the livingness of the polymerization.

The process of conventional living cationic polymerization is schematically illustrated in Fig. 10.9a. An additive is added to form an equilibrium between the active species and the dormant species. In this case, not all the polymer chains propagate at the same time, but only a small number of chains are propagating, whereas many of the polymer chains are resting as dormant species. These species occasionally become active and propagate. In this manner, the polymer chain extends slowly as a whole, causing a narrower distribution of chain lengths. When

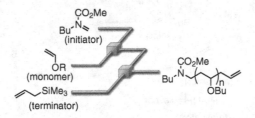

Fig. 10.7 Termination with allyltrimethylsilane

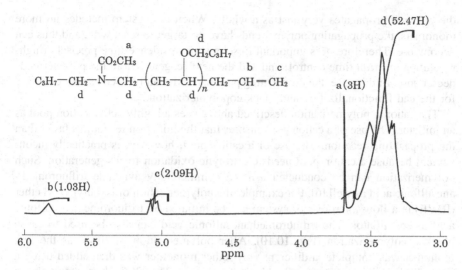

Fig. 10.8 ^{1}H NMR spectrum of the polymer obtained by termination with allyltrimethylsilane

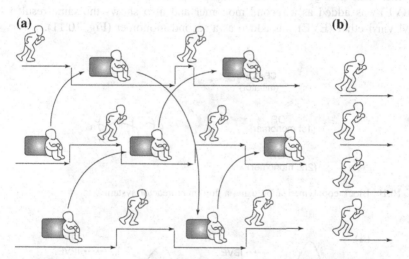

Fig. 10.9 Schematic illustrations of **a** conventional living polymerization and **b** living polymerization using a flow microreactor (flash polymerization)

the system includes no more monomers, the propagating polymer ends have no target to react with. However, most of the propagating polymer ends form dormant species and thus do not decompose appreciably. When another monomer is added to the system, the polymer slowly propagates to form a block copolymer.

In contrast, the flash polymerization using a flow microreactor system involves no additive and thus does not form an equilibrium between the active species and the dormant species (Fig. 10.9b). All the polymer ends form active species. Thus, the polymer propagates very fast as a whole. When the system includes no more monomers, the propagating polymer ends have no target to react with, and thus can decompose. Therefore, it is important to set a short residence time precisely (high-resolution reaction time control) and add the next reagent before the polymer ends decompose. In this case, the propagating polymer ends can be used as active species for the end functionalization and block copolymerization.

The cationic polymerization described above uses a highly active cation pool as an initiator. The use of a cation pool ensures that the initiation reaction is faster than the propagating reaction. The use of a cation pool, however, is practically incon-venient because a cation pool needs electrolytic oxidation for its generation. Such polymerization can be conducted using a commercially available trifluoromethanesulfonic acid as well [6]. For example, the polymerization of isobutyl vinyl ether (IBVE) in a flow microreactor system can be initiated by trifluoromethanesulfonic acid as an initiator. The trifluoromethanesulfonic acid can also be used to cause block copolymerization (Fig. 10.10). After polymerization of IBVE as the first monomer was complete, a different vinyl ether monomer was then added using a second micromixer to produce a block copolymer. The GPC chart shows that a polymer with a larger molecular weight was obtained when n-butyl vinyl ether (NBVE) was added as a second monomer and also shows the same result when ethyl vinyl ether (EVE) was added as a second monomer (Fig. 10.11).

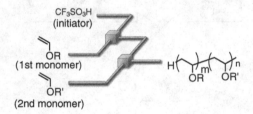

Fig. 10.10 Block copolymerization using a flow microreactor system

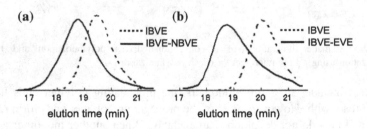

Fig. 10.11 GPC traces of the copolymers. **a** IBVE-NBVE and **b** IBVE-EVE

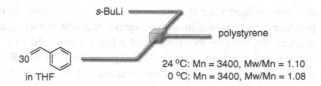

Fig. 10.12 Flash anionic polymerization of styrene using a flow microreactor system

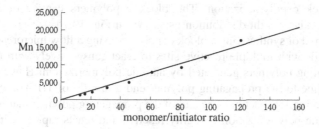

Fig. 10.13 Relationship between the molecular weight and the monomer/initiator ratio in flash anionic polymerization using a flow microreactor

10.5 Flash Anionic Polymerization

The flash method using a flow microreactor system is also useful for anionic poly-merization [7, 8]. Typically, when anionic polymerization of styrene is conducted using a batch macroreactor, the end of the propagating polymer is unstable, although the stability depends on the nature of the solvent. The polymerization in a polar solvent such as tetrahydrofuran (THF) needs to be conducted at an extremely low temperature of, for example, −78 °C. In a nonpolar solvent, the polymerization can be conducted at higher temperatures, but the reaction is slower and requires a longer reaction time. With these problems, anionic polymerization of styrene finds almost no industrial uses. Such problems can be solved by the flow microreactor system.

The flow microreactor system allows anionic polymerization of styrene in a polar solvent such as THF at 24 °C or 0 °C by using s-BuLi as an initiator (Fig. 10.12) [9]. The molecular weight increased in proportion to the ratio of the monomer to the initiator (with the molecular weight distribution of about 1.1), indicating that the end of the propagating polymer is active at around room tem-perature (Fig. 10.13).

The microreactor is useful not only for polymerization in a polar solvent but also for polymerization in a nonpolar solvent, such as cyclohexane. The polymerization in a nonpolar solvent can be conducted at 80 °C to produce a polymer with an extremely narrow distribution of molecular weights [10].[1]

---

[1]This paper also reported the anionic polymerization of styrenes in THF using a flow microreactor system.

An integrated flow microreactor system further allows the anionic living end of a propagating polymer to be usable in a subsequent reaction (space integration of reactions). For example, styrene monomer and *s*-BuLi are mixed in a first mixer (**M1**) and the anionic polymerization is conducted in the first reactor (**R1**), and then, chlorosilane is introduced from a second micromixer (**M2**) (Fig. 10.14). This can form a polymer with the propagating end quantitatively trapped by a silyl group with narrow molecular weight distribution.

The propagating polymer end may react with another styrene monomer to conduct block copolymerization. The block copolymers can be obtained with narrow molecular weight distribution as shown in Fig. 10.15.

This method of synthesizing a block copolymer using a flow microreactor allows an electrophile with multiple reactive sites to react consecutively with the ends of two propagating polymers generated by anionic polymerization. However, adding the electrophile to the propagating polymer end at a ratio of 1:1 would not easily lead the selective 1:1 reaction. The end of the propagating polymer that is generated by the anionic polymerization is highly reactive and reacts rapidly with the electrophile. This fast reaction has the problem of disguised chemical selectively in competitive consecutive reactions as discussed in Chap. 7. The selective reaction of the end of one propagating polymer with one molecule of electrophile normally requires the use of an excessively large amount of electrophile. The excess

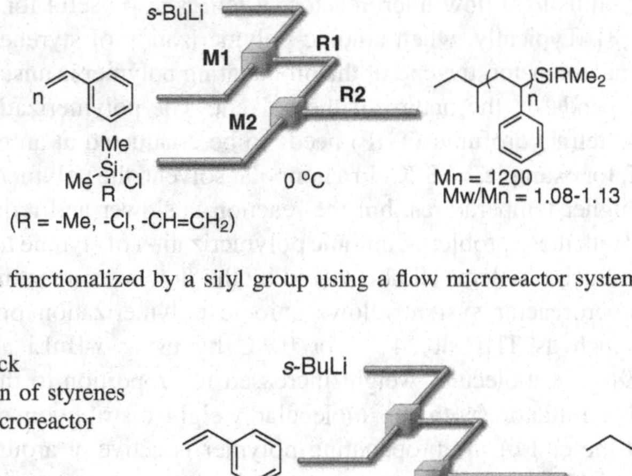

Fig. 10.14 End functionalized by a silyl group using a flow microreactor system

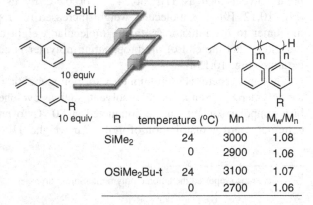

Fig. 10.15 Block copolymerization of styrenes using a flow microreactor system

R	temperature (°C)	Mn	M_w/M_n
SiMe$_2$	24	3000	1.08
	0	2900	1.06
OSiMe$_2$Bu-t	24	3100	1.07
	0	2700	1.06

electrophile thus needs to be removed before the polymer is allowed to react with the end of another propagating polymer generated by anionic polymerization. This disables continuous reaction with the next propagating polymer end. The flow microreactor system incorporating a micromixer enables selective 1:1 reaction by using 1 equiv. of electrophile with the propagating polymer end and thus allows the resulting polymer intermediate to react with the end of another propagating polymer without the need for separation and purification.

As shown in Fig. 10.16, the use of integrated flow microreactor system allows two propagating ends, which are generated by anionic polymerization of two different styrene monomers to react consecutively with dichlorodiorganosilane to form a block polymer. Two chlorine atoms attached to the silicon atom are consecutively replaced by the two polymer chains. Thus, a block copolymer having a silicon core can be easily synthesized.

Polymers of alkyl methacrylates are important materials that have many industrial uses. In anionic polymerization of an alkyl methacrylate, however, the initiator or the living end of the propagating polymer can undergo a nucleophilic reaction with the ester carbonyl group. With the need to suppress such side reactions, anionic polymerization of methacrylates has been more difficult than anionic polymerization of styrenes [11, 12]. The flow microreactor system allows anionic polymerization of methacrylates under the conditions that are easily accessible in industrial production (-30 °C or above) [13–15].

Anionic polymerization of alkyl methacrylates cannot be conducted using s-BuLi, which is used as an initiator of anionic polymerization of styrenes. To suppress the nucleophilic attack to the carbonyl group, anionic polymerization of alkyl methacrylate is usually initiated by 1,1-diphenylhexyllithium (DPHLi), which is prepared by the reaction of n-BuLi and 1,1-diphenylethene. This initiator has greater steric effects and is less reactive than s-BuLi, and thus is less likely to give a nucleophilic attack to the ester carbonyl group. Anionic polymerization of methyl

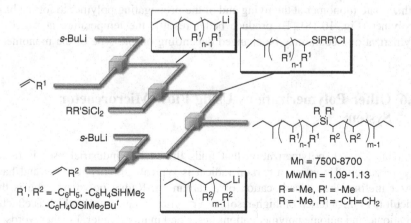

Fig. 10.16 Synthesis of block copolymers having a silicon core using an integrated flow microreactor system

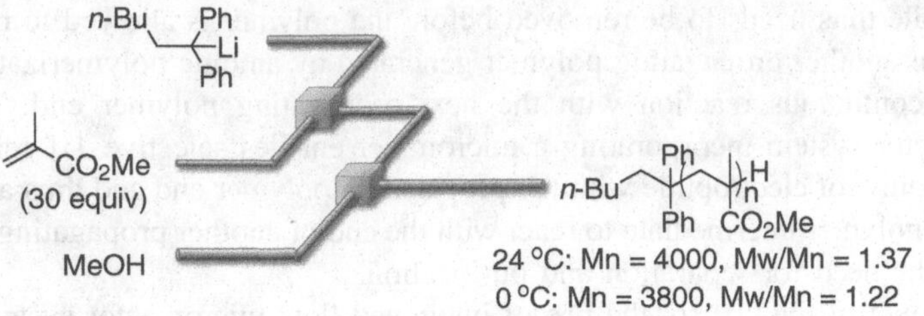

Fig. 10.17 Polymerization of methyl methacrylate using a flow microreactor

Fig. 10.18 Block copolymerization of alkyl methacrylates using a flow microreactor

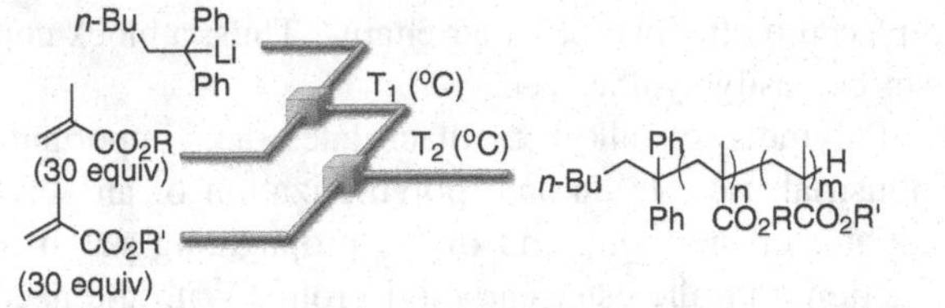

monomer 1	monomer 2	T^1 (°C)	T^2 (°C)	Mn	M_w/M_n
CO_2Bu-t	CO_2Bu-n	24	0	9500	1.16
	CO_2Me	24	-28	8400	1.15
CO_2Bu-n	CO_2Me	0	-28	9000	1.31

methacrylate (MMA) using this initiator can be conducted in the flow microreactor system to form a polymer with a relatively narrow distribution of molecular weights, without requiring extremely low-temperature environments (Fig. 10.17).

The integrated flow microreactor system may also be used to add another alkyl methacrylate monomer at the living end of the propagating polymer to form a block copolymer (Fig. 10.18). To produce better results, the temperature at which the polymerization occurs may be changed depending on the type of the monomer.

10.6 Other Polymerizations Using Flow Microreactor System

The chain-growth polymerization that finds the widest industrial uses is radical polymerization. However, a carbon radical is typically highly reactive and has a shorter lifetime than a carbocation or a carbanion. Radial polymerization is thus difficult to control based on high-resolution reaction time control, which is effective for cationic and anionic polymerizations described in this chapter. In other words, by using a simple flow microreactor technique radical polymerization cannot be living polymerization, in which the end of a propagating polymer remains active and is used in subsequent reactions or polymerization. The lifetime of a radical is normally too short for the present flow microreactor systems. However, the fast heat transfer of a flow microreactor system enables better control of the molecular weight or the molecular weight distribution than batch macroreactors [16–18]. To achieve living radical polymerization, an additive that reacts with a radical propagating end to form a dormant species should be added even using a flow microreactor [19–22].

Step-growth polymerization does not involve an unstable active species at the propagating polymer end. Thus, the molecular weight or the molecular weight distribution cannot be controlled, in principle, based on high-resolution reaction time control using a flow microreactor. Step-growth polymerization is typically not very fast and thus generates less heat per unit time than radical or other chain-growth polymerizations. The fast heat transfer of a flow microreactor system cannot be effective in such polymerizations. Although there are some advantage of flow microreactor systems for step-growth polymerization [23–25], the flow microreactor system is thus more effective for chain-growth polymerization.

References

1. M. Miyamoto, M. Sawamoto, T. Higashimura, Macromolecules **17**, 265 (1984)
2. T. Higashimura, M. Sawamoto, Adv. Polym. Sci. **62**, 49 (1984)
3. M. Sawamoto, Prog. Polym. Sci. **16**, 111 (1991)
4. A. Nagaki, K. Kawamura, S. Suga, T. Ando, M. Sawamoto, J. Yoshida, J. Am. Chem. Soc. **126**, 14702 (2004)
5. A. Nagaki, T. Iwasaki, K. Kawamura, D. Yamada, S. Suga, T. Ando, M. Sawamoto, J. Yoshida, Chem. Asian J. **3**, 1558 (2008)
6. T. Iwasaki, A. Nagaki, J. Yoshida, Chem. Commun. 1263 (2007)
7. A. Hirao, S. Loykulnant, T. Ishizone, Prog. Polym. Sci. **27**, 1399 (2002)
8. N. Hadjichristidis, M. Pitsikalis, S. Pispas, H. Iatrou, Chem. Rev. **101**, 3747 (2001)
9. A. Nagaki, Y. Tomida, J. Yoshida, Macromolecules **41**, 6322 (2008)
10. F. Wurm, D. Wilms, J. Klos, H. Lowe, H. Frey, Macromol. Chem. Phys. **209**, 1106 (2008)
11. C. Zune, R. Jérôme, Prog. Polym. Sci. **24**, 631 (1999)
12. D. Baskaran, Prog. Polym. Sci. **28**, 521 (2003)
13. A. Nagaki, Y. Tomida, A. Miyazaki, J. Yoshida, Macromolecule **42**, 4384 (2009)
14. A. Nagaki, A. Miyazaki, J. Yoshida, J. Macromolecules **43**, 8424 (2010)
15. A. Nagaki, A. Miyazaki, Y. Tomida, J. Yoshida, Chem. Eng. J. **167**, 548 (2011)
16. T. Iwasaki, J. Yoshida, Macromolecules **38**, 1159 (2005)
17. Z. Liu, Y. Lu, B. Yang, G. Luo, Ind. Eng. Chem. Res. **50**, 11853 (2011)
18. D. Parida, C.A. Serra, D.K. Garg, Y. Hoarau, R. Muller, M. Bouquey, Macromol. React. Eng. **8**, 597 (2014)
19. Y. Shen, S. Zhu, R. Pelton, Macromol. Rapid Commun. **21**, 956 (2000)
20. T. Wu, Y. Mei, J.T. Cabral, C. Xu, K.L. Beers, J. Am. Chem. Soc. **126**, 9880 (2004)
21. C. Rosenfeld, C. Serra, C. Brochon, G. Hadziioannou, Chem. Eng. Sci. **62**, 5245 (2007)
22. T. Fukuyama, Y. Kajihara, I. Ryu, A. Studer, Synthesis **44**, 2555 (2012)
23. S. Liu, C.H. Chang, Chem. Eng. Technol. **30**, 334 (2007)
24. D. Kessler, H. Löwe, P. Theato, Macromol. Chem. Phys. **210**, 807 (2009)
25. H. Seyler, D.J. Jones, A.B. Holmes, W.W.H. Wong, Chem. Commun. **48**, 1598 (2012)

Chapter 11
Outlook

In this book, we have learned that flow microreactor chemistry allows highly selective chemical transformations with the emphasis on the principle of high-resolution reaction time control and extremely fast micromixing. Flash chemistry using flow microreactors makes highly unstable short-lived reactive intermediates usable in a controlled manner and enables transformations that cannot be done in batch. Space integration of reactions makes multistep chemical synthesis much more efficient. Impact of flow microreactor chemistry in polymer synthesis is remarkable, because of superior controllability of molecular weight and molecular weight distribution.

We should keep in mind, however, that there are many other aspects of flow microreactor chemistry. For example, fast heat transfer because of high surface-to-volume ratio enables the performance of thermal reactions at high temperatures more efficiently. High surface-to-volume ratio is also beneficial in biphasic reactions or multiphasic reactions in which materials should be transferred through the interface. From a similar viewpoint, catalytic reactions using heterogeneous catalysts can enjoy high efficiency in flow microreactors. Similarly, the electrochemical reactions which occur on the surface of the electrodes should also be effective in flow microreactors. A short distance between the cathode and the anode lowers the cell voltage and sometimes enables the electrolysis without intentionally added electrolytes. Photochemical reactions in batch usually suffer from the problem of short distance of light penetration, which can be solved by using flow microphotochemical reactors. Books and review articles listed in the appendix and references cited therein would be helpful to understand the details of these features of flow microreactor chemistry.

Flow microreactor chemistry is still an ongoing challenging field of science and technology. The features of flow microreactor chemistry will hopefully work together to help meet the great demands of chemical synthesis and chemical production for the development of sustainable society in the near future.

© The Author(s) 2015
J. Yoshida, *Basics of Flow Microreactor Synthesis*,
SpringerBriefs in Molecular Science, DOI 10.1007/978-4-431-55513-1_11

Appendix

Books on Flow Microreactor Synthesis

1. W. Ehrfeld, *Microreaction Technology* (Springer, Berlin, 1998)
2. W. Ehrfeld, V. Hessel, H. Löwe, *Microreactors* (Wiley-VCH, Weinheim, 2000)
3. V. Hessel, S. Hardt, H. Löwe, *Chemical Micro Process Engineering* (Wiely-VCH, Weinheim, 2004)
4. V.M. Harik, L.-S. Luo, *Micromechanics and Nanoscale Effects: Mems, Multi-Scale Materials and Micro-Flows* (Kluwer Academic Pub, Boston, 2004)
5. V. Hessel, S. Hardt, H. Löwe, *Chemical Micro Process Engineering* (Wiley-VCH, Weinhem, 2004)
6. O. Geschke, H. Klank, P. Telleman, *Microsystem Engineering of Lab-on-a-Chip Devices* (Wiley-VCH, Weinheim, 2004)
7. G. Karniadakis, A. Beskok, N. Aluru, *Microflows and Nanoflows: Fundamentals and Simulation* (Springer, New York, 2005)
8. Y. Wang, J.D. Holladay, J. Holladay (eds.), *Microreactor technology and process intensification*, ACS Symposium Series, vol. 914, American Chemical Society, Washington, DC (2005)
9. J.-C. Chiao, D.N. Jamieson, L. Faraone, A.S. Dzurak, *Micro and nanotechnology: materials, processes, packaging, and systems II*. Proc. SPIE vol. 5650 (2005)
10. W.B.J. Zimmerman (ed.), *Microfluidics: History, Theory and Applications* (Springer, Wien, 2006)
11. P. Tabeling, S. Cheng, *Introduction to Microfluidics* (Oxford University Press, Oxford, 2006)
12. H. Bruus, *Theoretical Microfluidics* (Oxford Master Series in Physics) (Oxford University Press, Oxford, 2007)
13. T. Wirth (ed.), *Microreactors in Organic Synthesis and Catalysis* (Wiley-VCH, Weinheim, 2008)
14. J. Yoshida, *Flash Chemistry. Fast Organic Synthesis in Microsystems* (Wiley-Blackwell, Chichester, 2008)
15. T. Dietrich, *Principles and Applications of Chemical Microreactors* (Wiley, Hoboken, 2008)
16. N. Kockmann, *Transport Phenomena: In Micro Process Engineering* (Wiley, New York, 2008)
17. T. Wirth (ed.), *Microreactors in Organic Synthesis and Catalysis* (Wiley-VCH, Weinheim, 2008)
18. V. Hessel, A. Renken, J.C. Schouten, J. Yoshida (ed.), *Micro Process Engineering. A Comprehensive Handbook*, vol. 1, 2, and 3 (Wiley-VCH, Weinheim, 2009)

© The Author(s) 2015
J. Yoshida, *Basics of Flow Microreactor Synthesis*,
SpringerBriefs in Molecular Science, DOI 10.1007/978-4-431-55513-1

19. C. Wiles, P. Watts, *Micro Reaction Technology in Organic Synthesis* (CRC Press, Boca Raton, 2011)
20. W. Reschetilowski (ed.), *Microreactors in Preparative Chemistry* (Wiley-VCH, Weinheim, 2013)
21. T. Wirth (ed.), *Microreactors in Organic Synthesis and Catalysis. Second, Completely Revised and Enlarged Edition* (Wiley-VCH, Weinheim, 2013)

Review Articles on Flow Microreactor Synthesis

1. A. Shanley, Microreactors find new niches—tiny reactors and ancillary devices may work where large equipment would be too risky or costly. Chem. Eng. **30**, 1 March 1997
2. R.F. Service, Labs on a chip: miniaturization puts chemical plants where you want them. Science **282**, 400 (1998)
3. T. Von Zech, D. Hönicke, Microreaction technology: potentials and technical feasibility. Erdöl Erdgas Kohle **114**, 578 (1998)
4. W. Ehrfeld, V. Hassel, H. Lehr, Microreactors for chemical synthesis and biotechnology—current developments and future applications. Top. Curr. Chem. **194**, 233 (1998)
5. M. Harre, U. Tilstam, H. Weinmann, Breaking the new bottleneck: automated synthesis in chemical process research and development. Org. Process Res. Dev. **3**, 304 (1999)
6. H. Löwe, W. Ehrfeld, State-of-the-art in microreaction technology: concepts, manufacturing and applications. Electrochim. Acta **44**, 3679 (1999)
7. S.H. DeWitt, Microreactors for chemical synthesis. Curr. Opin. Chem. Biol. **3**, 350 (1999)
8. P. Fletcher, S. Haswell, Downsizing synthesis—investigate the benefits of performing chemical synthesis in microscale. Chem. Br. **38** (1999)
9. M. Freemantle, Downsizing chemistry—chemical analysis and synthesis on microchips promise a variety of potential benefits. Chem. Eng. News **22**, 27 (1999)
10. H. Gau, S. Herminghaus, P. Lenz, R. Lipowsky, Liquid morphologies on structured surfaces: from microchannels to microchips. Science **283**, 46 (1999)
11. S.J. Haswell, V. Skelton, Chemical and biochemical microreactors. Trends Anal. Chem. **19**, 389 (2000)
12. E. L'Hostis, P.E. Michel, G.C. Fiaccabrino, D.J. Strike, N.F. de Rooij, M. Koudelka-Hep, Microreactor and electrochemical detectors fabricated using Si and EPON SU-8. Sens. Actuators B **64**, 156 (2000)
13. G.M. Greenway, S.J. Haswell, D.O. Morgan, V. Skelton, P. Styring, The use of a novel microreactor for high throughput continuous flow organic synthesis. Sens. Actuators B **63**, 153 (2000)
14. M. Kakuta, F.G. Bessoth, A. Manz, Microfabricated devices for fluid mixing and their application for chemical synthesis. Chem. Rec. **1**, 395 (2001)
15. O. Wörz, K.-P. Jäckel, T. Richter, A. Wolf, Microreactors—a new efficient tool for reactor development. Chem. Eng. Technol. **24**, 138 (2001)
16. K. Benz, K.-P. Jäckel, K.-J. Regenauer, J. Schiewe, K. Drese, W. Ehrfeld, V. Hessel, H. Löwe, Utilization of micromixers for extraction processes. Chem. Eng. Technol. **24**, 11 (2001)
17. O. Wörz, K.P. Jäckel, T. Richter, A. Wolf, Microreactors, a new efficient tool for optimum reactor design. Chem. Eng. Sci. **56**, 1029 (2001)
18. K.F. Jensen, Microreaction engineering—is small better? Chem. Eng. Sci. **56**, 293 (2001)
19. S.J. Haswell, R.J. Middleton, B. O'Sullivan, V. Skelton, P. Watts, P. Styring, The application of micro reactors to synthetic chemistry. Chem. Commun. **5**, 391 (2001)
20. A. Kirschning, H. Monenschein, R. Wittenberg, Functionalized polymers—emerging versatile tools for solution-phase chemistry and automated parallel synthesis. Angew. Chem. Int. Ed. **40**, 650 (2001)

21. S.J. Haswell, R.J. Middleton, B. O'Sullivan, V. Skelton, P. Watts, P. Styring, The application of micro reactors to synthetic chemistry. Chem. Commun. **5**, 391 (2001)

22. A. de Mello, R. Wooton, But what is it good for? applications of microreactor technology for the fine chemical industry. Lab Chip **2**, 7N–13N (2002)

23. V. Hessel, H. Löwe, T. Stange, Micro chemical processing at IMM–from pioneering work to customer-specific services. Lab Chip **2**, 14N–21N (2002)

24. B.M. Stone, A. de Mello, Life, the universe and microfluids. Lab Chip **2**, 58N–64N (2002)

25. P.D.I. Fletcher, S.J. Haswell, E. Pombo-Villar, B.H. Warrington, P. Watts, S.Y.F. Wong, X. Zhang, Micro reactors: principles and applications in organic synthesis. Tetrahedron **58**, 4735 (2002)

26. S.V. Ley, I.R. Baxendale, New tools and concepts for modern organic synthesis. Nat. Rev. Drug Discovery **1**, 573 (2002)

27. B.M. Stone, A.J. de Mello, Life, the universe and Microfluidics. Lab Chip **2**, 58N (2002)

28. V. Hessel, H. Löwe, T. Stange, Micro chemical processing at IMM-from pioneering work to customer-specific services. Lab Chip **2**, 14N (2002)

29. A. de Mello, R. Wootton, But what is it good for? applications of microreactor technology for the fine chemical industry. Lab Chip **2**, 7N (2002)

30. T. Schwalbe, V. Autze, G. Wille, Chemical synthesis in microreactors. Chimia **56**, 636 (2002)

31. P.-A. Auroux, D. Iossifidis, D.R. Reyes, A. Manz, Micro total analysis systems. 2. Analytical standard operations and applications. Anal. Chem. **74**, 2637 (2002)

32. D.R. Reyes, D. Iossifidis, P.-A. Auroux, A. Manz, Micro total analysis systems. 1. Introduction, theory, and technology. Anal. Chem. **74**, 2623 (2002)

33. A.J. de Mello, Miniaturization. Anal. Bioanal. Chem. **372**, 12 (2002)

34. S.J. Haswell, P. Watts, Green chemistry: synthesis in micro reactors. Green Chem. **5**, 240 (2003)

35. R.F. Ismagilov, Integrated microfluidic systems. Angew. Chem. Int. Ed. **42**, 4130 (2003)

36. G. Jas, A. Kirschning, Continuous flow techniques in organic synthesis. Chem. Eur. J **9**, 5708 (2003)

37. P. Watts, S.J. Haswell, Microfluidic combinatorial chemistry. Curr. Opin. Chem. Biol. **7**, 380 (2003)

38. P. Watts, S.J. Haswell, Continuous flow reactors for drug discovery. Drug Discovery Today **8**, 586 (2003)

39. S.J. Haswell, P. Watts, Green chemistry: synthesis in micro reactors. Green Chem. **5**, 240 (2003)

40. P. Watts, S.J. Haswell, Microfluidic combinatorial chemistry. Curr. Opin. Chem. Biol. **7**, 380 (2003)

41. V. Hessel, H. Löwe, Microchemical engineering: components, plant concepts, user acceptance-part I. Chem. Eng. Technol. **26**, 13 (2003)

42. V. Hessel, H. Löwe, Microchemical engineering: components, plant concepts, user acceptance-part II. Chem. Eng. Technol. **26**, 391 (2003)

43. V. Hessel, H. Löwe, Microchemical engineering: components, plant concepts, user acceptance-part III. Chem. Eng. Technol. **26**, 531 (2003)

44. G. Jas, A. Kirschning, Continuous flow techniques in organic synthesis. Chem. Eur. J. **9**, 5708 (2003)

45. H. Song, J.D. Tice, R.F. Ismagilov, A microfluidic system for controlling reaction networks in time. Angew. Chem. Int. Ed. **42**, 767 (2003)

46. S.A. Agnihotri, N.N. Mallikarjuna, T.M. Aminabhavi, Recent advances on chitosan-based micro- and nanoparticles in drug delivery. J. Control Release **100**, 5 (2004)

47. A. de Mello, J. de Mello, Microscale reactors: nanoscale products. Lab Chip **4**, 11N (2004)

48. Q. Fang, Sample introduction for microfluidic systems. Anal. Bioanal. Chem. **378**, 49 (2004)

49. P. Gould, Microfluidics realizes potential. Mater. Today **7**, 48 (2004)

50. S. Hasebe, Design and operation of micro-chemical plants—bridging the gap between nano, micro and macro technologies. Comput. Chem. Eng. **29**, 57 (2004)

51. J.D. Hollady, Y. Wang, E. Jones, Review of developments in portable hydrogen production using microreactor technology. Chem. Rev. **104**, 4767–4790 (2004)

52. K. Jahnisch, V. Hessel, H. Löwe, M. Baerns, Chemistry in microstructured reactors. Angew. Chem. Int. Ed. Engl. **43**, 406 (2004)

53. Y. Kikutani, T. Kitamori, Micro-flow reaction systems for combinatorial syntheses. Macromol. Rapid Commun. **25**, 158 (2004)

54. P. Löb, H. Löwe, V. Hessel, Fluorinations, chlorinations and brominations of organic compounds in micro reactors. J. Fluorine Chem. **125**, 1677 (2004)

55. N. Minc, J.-L. Viovy, Microfluidique et applications biologiques: enjeux et tendances. C. R. Phys. **5**, 565 (2004)

56. H. Pennemann, V. Hessel, H. Löwe, Chemical microprocess technology—from laboratory-scale to production. Chem. Eng. Sci. **59**, 4789 (2004)

57. C.S. Peyratout, L. Dahne, Tailor-made polyelectrolyte microcapsules: from multilayers to smart containers. Angew. Chem. Int. Ed. **43**, 3762 (2004)

58. A. Renken, P.J. Baselt, M. Matlosz, Special issue: microreaction technolgy. Chem. Eng. J. **101**, 1 (2004)

59. T. Schwalbe, V. Autze, M. Hohmann, W. Stirner, Novel innovation systems for a cellular approach to continuous process chemistry from discovery to market. Org. Process Res. Dev. **8**, 440 (2004)

60. X. Zhang, S. Stefanick, F.J. Villani, Application of microreactor technology in process development. Org. Process Res. Dev. **8**, 455 (2004)

61. J. Yoshida, Flash chemistry using electrochemical method and microsystems. Chem. Commun. **36**, 4509 (2005)

62. D. Belder, Microfluidics with droplets. Angew. Chem. Int. Ed. Engl. **44**, 3521 (2005)

63. G.N. Doku, W. Verboom, D.N. Reinhoudt, A. van den Berg, On-microchip multiphase chemistry—a review of microreactor design principles and reagent contacting modes. Tetrahedron **61**, 2733 (2005)

64. V. Hessel, H. Löwe, F. Schönfeld, Micromixers—a review on passive and active mixing principles. Chem. Eng. Sci. **60**, 2479–2501 (2005)

65. L. Kiwi-Minsker, A. Renken, Microstructured reactors for catalytic reactions. Catal. Today **110**, 2 (2005)

66. P. Watts, S.J. Haswell, The application of micro reactors for organic synthesis. Chem. Soc. Rev. **34**, 235 (2005)

67. M. Brivio, W. Verboom, D.N. Reinhoudt, Miniaturized continuous flow reaction vessels: influence on chemical reactions. Lab Chip **6**, 329 (2006)

68. A.J. de Mello, Control and detection of chemical reactions in microfluidic systems. Nature **442**, 394 (2006)

69. A. Kirschning, W. Solodenko, K. Mennecke, Combining enabling techniques in organic synthesis: continuous flow processes with heterogenized catalysts. Chem. Eur. J. **12**, 5972 (2006)

70. J. Kobayashi, Y. Mori, S. Kobayashi, Multiphase organic synthesis in microchannel reactors. Chem. Asian J. **1**, 22 (2006)

71. H. Song, D.L. Chen, R.F. Ismagilov, Reactions in droplets in microfluidic channels. Angew. Chem. Int. Ed. **45**, 7336 (2006)

72. G. Sui, H.-R. Tseng, Reactions in hand. Nano Today **1**, 6 (2006)

73. J. Wang, G. Sui, V.P. Mocharla, R.J. Lin, M.E. Phelps, H.C. Kolb, H.-R. Tseng, Integrated microfluidics for parallel screening of an in situ click chemistry library. Angew. Chem. Int. Ed. **45**, 5276 (2006)

74. G.M. Whitesides, The origins and the future of microfluidics. Nature **442**, 368 (2006)

75. B. Ahmed-Omer, J.C. Brandt, T. Wirth, Advanced organic synthesis using microreactor technology. Org. Biomol. Chem. **5**, 733 (2007)

76. B.P. Mason, K.E.S. Price, J.L. Steinbacher, A.R. Bdgdan, D.T. McQuade, Greener approaches to organic synthesis using microreactor technology. Chem. Rev. **107**, 2300 (2007)
77. P.L. Mills, D.J. Quiram, J.F. Ryley, Microreactor technology and process miniaturization for catalytic reactions—A perspective on recent developments and emerging technologies. Chem. Eng. Sci. **62**, 6992 (2007)
78. P. Watts, C. Wiles, Recent advances in synthetic micro reaction technology. Chem. Commun. **5**, 443 (2007)
79. V. Hessel, D. Kralischc, U. Krtschilb, Sustainability through green processing – novel process windows intensify micro and milli process technologies. Energy Environ. Sci. **1**, 467 (2008)
80. N. Kockmann, M. Gottsponer, B. Zimmermann, D.M. Roberge, Enabling continuous-flow chemistry in microstructured devices for pharmaceutical and fine-chemical production. Chem. Eur. J. **14**, 7470 (2008)
81. D.M. Roberge, B. Zimmermann, F. Rainone, M. Gottsponer, M. Eyholzer, N. Kockmann, Microreactor technology and continuous processes in the fine chemical and pharmaceutical industry: is the revolution underway? Org. Process Res. Dev. **12**, 905 (2008)
82. I. Ryu, T. Fukuyama, M. Rahman, M. Sato, Adventures in inner space: microflow systems for practical organic synthesis. Synlett **2**, 151 (2008)
83. D. Wilms, J. Klos, H. Frey, Microstructured reactors for polymer synthesis: a renaissance of continuous flow processes for tailor-made macromolecules? Macromol. Chem. Phys. **209**, 343 (2008)
84. J. Yoshida, A. Nagaki, T. Yamada, Flash chemistry: fast chemical synthesis by using microreactors. Chem. Eur. J. **14**, 7450 (2008)
85. A.M. Elizarov, Microreactors for radiopharmaceutical synthesis. Lab Chip **9**, 1326 (2009)
86. K. Geyer, T. Fustafsson, P.H. Seeberger, Developing continuous-flow microreactors as tools for synthetic chemists. Synlett **15**, 2382 (2009)
87. R.L. Hartman, K.F. Jensen, Microchemical systems for continuous-flow synthesis. Lab Chip **9**, 2495 (2009)
88. V. Hessel, Novel process windows - gate to maximizing process intensification via flow chemistry. Chem. Eng. Technol. **32**, 1655 (2009)
89. W.Y. Lin, Y. Wang, S. Wang, H.R. Tseng, Integrated microfluidic reactors. Nano Today **4**, 470 (2009)
90. K. Tanaka, K. Fukase, Renaissance of traditional organic reactions under microfluidic conditions: a new paradigm for natural products synthesis. Org. Process Res. Dev. **13**, 983 (2009)
91. J. Yoshida, Organic electrochemistry, microreactors, and their synergy. Interface **18**, 40 (2009, summer)
92. C.G. Frost, L. Mutton, Heterogeneous catalytic synthesis using microreactor technology. Green Chem. **12**, 1687 (2010)
93. T. Illg, P. Lob, V. Hessel, Flow chemistry using milli- and microstructured reactors-from conventional to novel process windows. Bioorg. Med. Chem. **18**, 3707 (2010)
94. S.V. Ley, The changing face of organic synthetism. Tetrahedron **66**, 6270 (2010)
95. S. Marre, K.F. Jensen, Synthesis of micro and nanostructures in microfluidic systems. Chem. Soc. Rev. **39**, 1183 (2010)
96. J.P. McMullen, K.F. Jensen, Integrated microreactors for reaction automation: new approaches to reaction development. Annu. Rev. Anal. Chem. **3**, 19 (2010)
97. T. Razzaq, C.O. Kappe, Continuous flow organic synthesis under high-temperature/pressure conditions. Chem. Asian J. **5**, 1274 (2010)
98. F.E. Valera, M. Quaranta, A. Moran, J. Blacker, A. Armstrong, J.T. Cabral, D.G. Blackmond, The flow's the thing or is it? Assessing the merits of homogeneous reactions in flask and flow. Angew. Chem. Int. Ed. **49**, 2478 (2010)

99. D. Webb, T.F. Jamison, Continuous flow multi-step organic synthesis. Chem. Sci. **1**, 675 (2010)
100. J. Yoshida, Flash chemistry: flow microreactor synthesis based on high-resolution reaction time control. Chem. Rec. **10**, 332 (2010)
101. R.L. Hartman, J.P. McMullen, K.F. Jensen, Deciding whether to go with the flow: evaluating the merits of flow reactors for synthesis. Angew. Chem. Int. Ed. **50**, 7502 (2011)
102. M. Rasheed, T. Wirth, Intelligent microflow: development of self-optimizing reaction systems. Angew. Chem. Int. Ed. **50**, 357 (2011)
103. J. Wegner, S. Ceylan, A. Kirschning, Ten key issues in modern flow chemistry. Chem. Commun. **47**, 4583 (2011)
104. C. Wiles, P. Watts, Recent advances in micro reaction technology. Chem. Commun. **47**, 6512 (2011)
105. J. Yoshida, H. Kim, A. Nagaki, Green and sustainable chemical synthesis using flow microreactors. ChemSusChem **4**, 331 (2011)
106. J. Yoshida, K. Saito, T. Nokami, A. Nagaki, Space integration of reactions: an approach to increase the capability of organic synthesis. Synlett **2011**, 1189 (2011)
107. A.A. Desai, Overcoming the limitations of lithiation chemistry for organoboron compounds with continuous processing. Angew. Chem. Int. Ed. **51**, 9223 (2012)
108. A. Kirschning, L. Kupracz, J. Hartwig, New synthetic opportunities in miniaturized flow reactors with inductive heating. Chem. Lett. **41**, 562 (2012)
109. C. Wiles, P. Watts, Continuous flow reactors: a perspective. Green Chem. **14**, 38 (2012)
110. A. Nagaki, J. Yoshida, Controlled polymerization in flow microreactor systems. Adv. Polym. Sci. **259**, 1 (2013)
111. L. Malet-Sanz, F. Susanne, Continuous flow synthesis. a pharma perspective. J. Med. Chem. **55**, 4062 (2012)
112. I.R. Baxendale, The integration of flow reactors into synthetic organic chemistry. J. Chem. Technol. Biotechnol. **88**, 519 (2013)
113. D.T. McQuade, P.H. Seeberger, Applying flow chemistry: methods, materials, and multistep synthesis. J. Org. Chem. **78**, 6384 (2013)
114. J.C. Pastre, D.L. Browne, S.V. Ley, Flow chemistry syntheses of natural products. Chem. Soc. Rev. **42**, 8849 (2013)
115. J. Yoshida, A. Nagaki, D. Yamada, Continuous flow synthesis. Drug Discov. Today Technol. **10**, e53 (2013)
116. J. Yoshida, Y. Takahashi, A. Nagaki, Flash chemistry: flow chemistry that cannot be done in batch. Chem Commun. **49**, 9896 (2013)
117. Q. Tu, L. Pang, Y. Zhang, M. Yuan, J. Wang, D. Wang, W. Liu, J. Wang, Microfluidic device: a miniaturized platform for chemical reactions. Chin. J. Chem. **31**, 304 (2013)
118. K.S. Elvira, X. Casadevall i Solvas, R.C.R. Wootton, A.J. deMello, The past, present and potential for microfluidic reactor technology in chemical synthesis. Nat. Chem. **5**, 905 (2013)
119. T. Fukuyama, T. Totoki, I. Ryu, Carbonylation in microflow: close encounters of CO and reactive species. Green Chem. **16**, 2042 (2014)
120. Y. Su, N.J.W. Straathof, V. Hessel, T. Noël, Photochemical transformations accelerated in continuous-flow reactors: basic concepts and applications. Chem. Eur. J. **20**, 10562 (2014)
121. A. Manvar, A. Shah, Subtle Mitsunobu couplings under super-heating: the role of high-throughput continuous flow and microwave strategies. Org. Biomol. Chem. **12**, 8112 (2014)
122. C. Len, R. Luque, Continuous flow transformations of glycerol to valuable products: an overview. Sustain. Chem. Processes **2**, 1 (2014)
123. A.A. Kulkarni, Continuous flow nitration in miniaturized devices. Beilstein J. Org. Chem. **10**, 405 (2014)

Printed in the United States
By Bookmasters